Returning People to the Moon After Apollo
Will It Be Another Fifty Years?

Pat Norris

Returning People to the Moon After Apollo

Will It Be Another Fifty Years?

Published in association with
Praxis Publishing
Chichester, UK

Pat Norris
Byfleet, Surrey, UK

SPRINGER-PRAXIS BOOKS IN SPACE EXPLORATION

Springer Praxis Books
ISBN 978-3-030-14914-7   ISBN 978-3-030-14915-4 (eBook)
https://doi.org/10.1007/978-3-030-14915-4

© Springer Nature Switzerland AG 2019

This work is subject to copyright. All rights are reserved by the Publisher, whether the whole or part of the material is concerned, specifically the rights of translation, reprinting, reuse of illustrations, recitation, broadcasting, reproduction on microfilms or in any other physical way, and transmission or information storage and retrieval, electronic adaptation, computer software, or by similar or dissimilar methodology now known or hereafter developed.

The use of general descriptive names, registered names, trademarks, service marks, etc. in this publication does not imply, even in the absence of a specific statement, that such names are exempt from the relevant protective laws and regulations and therefore free for general use.

The publisher, the authors, and the editors are safe to assume that the advice and information in this book are believed to be true and accurate at the date of publication. Neither the publisher nor the authors or the editors give a warranty, express or implied, with respect to the material contained herein or for any errors or omissions that may have been made. The publisher remains neutral with regard to jurisdictional claims in published maps and institutional affiliations.

Cover design: Jim Wilkie

This Springer imprint is published by the registered company Springer Nature Switzerland AG
The registered company address is: Gewerbestrasse 11, 6330 Cham, Switzerland

*For Mealla, Ines and Éabha
bright stars for the 21st century*

Contents

Acknowledgements vii

1 **Why Did the United States Send Men to the Moon in the 1960s?** 1

2 **The American Rocket** 9

3 *Apollo 11* **– Getting There** 18

4 **The Eagle's Journey** 37

5 **American Knowhow** 57

6 **After** *Apollo 11* 68

7 **Legacy of Apollo** 74

8 **The Other Competitor in the Race** 83

9 **Why Is It Taking So Long to Return to the Moon?** 121

10 **When Will the United States Go Back?** 143

11 **China, the Communist Challenger** 178

12 **Russia and the Rest** 192

13 **Conclusions** 210

Glossary 215

Index 226

Acknowledgements

I am grateful to Bill Barry of NASA, and David Parker and Karl Berquist of the European Space Agency for very helpful discussions and advice. I also thank Brian Harvey for help and advice about getting suitable images. Thanks also to Asif Siddiqi and Elston Hill for their prompt agreement to use their images. And above all thanks to my very patient wife, who had to put up with what amounted to postponement of my retirement while I wrote this book.

1

Why Did the United States Send Men to the Moon in the 1960s?

The May 1961 decision to send men to the Moon was made at least in part for short-term political reasons. President John F. Kennedy wanted to restore his ratings following several weeks of relentlessly bad press, and the promise of a Moon landing seemed the best way to get the media and the public back on his side.

In November 1960 Kennedy had won one of the closest presidential elections in history. The results in two states in particular, Illinois and Texas, were crucial, and questioned by his opponents. If his Republican opponent Richard M. Nixon had carried both of those states he and not Kennedy would have won the election. Later legal challenges to the results failed. To his credit, Nixon accepted the election result immediately and urged his supporters to do likewise. As is still the case, the official result was based on counting the electoral votes of each state, rather than on the total votes received nationwide. As it so happens Kennedy received a slightly larger number of votes in total across the country,[1] which helped to legitimize his victory. Examples of a U. S. president being elected with a smaller total national vote than his opponent include President Donald Trump in 2016 with nearly 3 million votes less than Hilary Clinton, and President George W Bush in 2000 with half a million votes less than Al Gore.

Taking up office in January 1961, Kennedy exhibited great charisma and intelligence, charm and style. Starting with his inauguration address on the steps of the Capitol building, his speech writers provided him with stirring quotes such as the call to arms to his fellow Americans, to "ask not what your country can do for

---

[1] About 100,000 more out of a total of 68.8 million votes cast, a margin of about one sixth of one percent. As an example of the claims of fraud, Fannin County in Texas had 4,895 registered voters but recorded 6,138 votes cast, of which three quarters were for Kennedy. There were reports of Republican voter fraud, too, so in the end the election result was broadly supported.

2 Why Did the United States Send Men to the Moon in the 1960s?

you – ask what you can do for your country." In March he established the Peace Corps to provide the mechanism through which young Americans could "serve their country and the cause of peace by living and working in the developing world." But April 1961 was a difficult month for Kennedy.

The first blow came on April 12, when the Soviet Union launched the first man into space, Yuri Gagarin (see Fig. 1.1). The Soviets had taken the lead in space four years earlier with the launch of the first manmade object to orbit Earth, *Sputnik-1,* on October 4, 1957. Ironically Kennedy had benefited from the public outcry against the Soviet Union for dominating in the space race over the United States. He had criticized the policies of the previous U. S. president, Dwight D. Eisenhower ("Ike") that had allowed the Soviet Union to get a jump in the development of long range missiles and space satellites.

Fig. 1.1. April 12, 1961. Yuri Gagarin on his way to the launch site of his historic trip into space. At 9:07 a. m. Moscow time his Vostok rocket blasted off, placing his *Vostok-1* spaceship in orbit and making him the first human to orbit Earth. On its return, the spacecraft was not designed to land softly, so Gagarin (and other early Soviet cosmonauts) had to eject from the capsule at 23,000 ft (7 km) and make the final part of the descent by parachute; the capsule also landed via parachute nearby but at a speed that created a sizeable crater. (Illustration courtesy of ESA.)

Although the United States launched a satellite into space four months later, the Soviets continued to astound the world with space spectaculars, including launching much larger satellites than America could manage, sending the first probe to reach the Moon (September 1959), taking the first photographs of the hidden far side of the Moon (October 1959), sending the first animal into space (Laika the dog, November 1957) and sending the first animals to be recovered from space (August 1960). The Soviets had a very large number of failed launch attempts, but these were kept secret, so all that the public saw was the successes. By comparison

American launches took place in a full blaze of media attention, so that in comparison with the apparently infallible Soviet space technology, each American failure seemed to magnify the Soviet lead. In actual fact the Soviets suffered more failed launches and fewer successful launches than the United States in this period, but this only became known in the 1990s.[2]

The handsome, smiling face of Yuri Gagarin on the front page of the world's newspapers dulled some of the shine on the glittering start to Kennedy's presidency. Worse was to come.

President Eisenhower had left Kennedy with the headache of an anti-American government in Cuba, just 100 miles south of Florida. He had also left behind a group of Cuban exiles who were itching to invade Cuba and overthrow its government. Armed and trained by the CIA, 1,400 exiles set out from their base in Guatemala and reached the shores of Cuba on April 17 at the Bay of Pigs. Kennedy had vetoed the use of American ships and planes[3] to support the invasion, even though such support had been part of the original Eisenhower-era plan. The forces of Cuban leader Fidel Castro were deployed rapidly, surrounding the invaders who surrendered after three days. Castro's position as Cuban leader was greatly strengthened by these events, while Kennedy was made to look like an old-fashioned and ineffective imperialist.

Things improved in May. On May 5th Alan Shepard became the first American in space. His 15-minute Mercury flight took him 116 miles high – but not into orbit. His *Freedom-7* capsule landed safely in the Atlantic Ocean as planned, about 300 miles downrange from the launch site at Cape Canaveral, Florida (see Fig. 1.2).

From Kennedy's perspective the most interesting part of this was the enormous and positive reaction from the American public. Shepard became a hero overnight, attracting huge crowds wherever he went and generating positive publicity in the media. This public appetite for an American space hero strengthened Kennedy's tentative decision to announce that the United States would send a man to the Moon.

President Eisenhower had been reluctant to fund a human spaceflight program. He was, however, very keen to fund satellites carrying cameras rather than humans, and had begun such a program in 1954, more than three years *before* Sputnik. After Sputnik, Ike had resisted the demands of space enthusiasts to fund human space travel, but had reluctantly agreed to the relatively modest Mercury program. He decided to support the military funding for reconnaissance satellites by

---

[2] In 1958 the Soviets had 6 unpublicized launch failures and only one success, while the United States had ten failed launches and seven successes. The "score" should have been recognized as U. S. 7, Soviets 1, but because of the lack of information about the Soviet failures plus the media's tendency to highlight disasters over successes, the public perception was U. S. 10 failures, Soviets 1 success.

[3] A small number of U. S.-supplied planes did bomb Cuban airfields on April 16th.

4 Why Did the United States Send Men to the Moon in the 1960s?

Fig. 1.2. A U. S. Marine helicopter recovery team hoists astronaut Alan Shepard from his Mercury capsule after a successful flight and splashdown in the Atlantic Ocean. (Illustration courtesy of NASA.)

creating the National Aeronautics and Space Administration (NASA) to deal with both human spaceflight and scientific space initiatives.[4]

Kennedy was not necessarily a space enthusiast. A few years earlier, while still in the U. S. Senate, he and his brother Robert (Bobby) had treated with "good natured scorn" an attempt by Boston scientists to win their support for a U. S. spaceflight program. The Kennedys felt that "all rockets were a waste of money" [1], p. 61.

---

[4] Ike's enthusiasm for reconnaissance satellites was due to the lack of knowledge in the West about the secretive Soviet Union's military capabilities, and the fear of a surprise attack such as the Japanese had sprung on America less than 20 years earlier. The *Sputnik* and then Gagarin surprises strengthened the argument for better knowledge about Soviet rocketry that would be (and eventually was) provided by reconnaissance satellites.

Why Did the United States Send Men to the Moon in the 1960s?

The media backlash in the wake of Gagarin's April 12 flight began to change that. The next day Kennedy asked his adviser, Teddy Sorensen, to look at the space program options available to the United States. The day after that Sorensen held a strategy meeting with four colleagues: NASA Administrator James Webb, Webb's Deputy Hugh Dryden, Kennedy's Special Adviser for Science & Technology Jerome Wiesner, and the Budget Director David Bell. After a lengthy meeting they concluded that the Soviets would continue to lead in space for five years or so due to the greater power of their rockets. America needed to jump to the next generation of technology before it could outperform the Soviets.

The existing Soviet rockets meant that they would probably be the first to place spacecraft with two or even three astronauts into orbit, and perhaps the first to send humans around the Moon (but not land). With a new generation of rockets, in due course the United States could take the lead and by then the only significant objective left would be to land people on the Moon.

NASA had been analyzing larger rockets and their use in landing people on the Moon for some time. The subject of human spaceflight had been promoted for a decade or more by the *eminence gris* of the U. S. space program, Wernher von Braun (see Fig. 1.3). Having virtually invented long range rocketry in Germany during World War II, he now led the NASA rocket development activities at its Marshall Space Flight Center in Huntsville, Alabama, where his design for a massive new rocket was being worked out. He was aided inside NASA headquarters in Washington, D. C., by another European immigrant, Austrian-born George Low, whose particular hobbyhorse was to land men on the Moon.

On the face of it, the arguments for sending humans to explore the Moon were weak. Wiesner (the Science Adviser) had remarked that human spaceflight cost an order of magnitude more than robotic (unmanned) spacecraft.[5] However the political benefits of humans in space now began to outweigh the added cost.

Three months before Sorensen's April meeting (and just before Kennedy's inauguration), NASA had produced an outline of how to send people to the Moon, coming up with a schedule that saw a lunar landing taking place in 1969 or 1970. In March Kennedy had approved funds to continue work on von Braun's giant rocket, but he had refused to fund the design of the Apollo spacecraft that would be placed on top of the rocket. At that point, while the president was still of two minds about human spaceflight, Vice President Lyndon B Johnson was very much in favor of it. He used his legendary persuasive skills to drum up support in Congress and in the media for an ambitious U. S. space initiative.

---

[5] Humans need oxygen, water, food, heat, 8 hours sleep, etc. Machines need some electricity. Machines were getting smaller as the electronic revolution took off; men weren't.

6 Why Did the United States Send Men to the Moon in the 1960s?

Fig. 1.3. Walt Disney *(left)* visiting Wernher von Braun at his army missile base in Huntsville, Alabama, in 1954. In addition to managing the most advanced rocket development center in America, von Braun had helped during the 1950s to popularize the concept of space travel to the American public, working with Disney Studios as a technical director and making three films about space exploration for television. (Illustration courtesy of NASA.)

Thanks to the work by von Braun and others, NASA now had a certain amount of data to underpin a decision about sending men to the Moon, and this was enough to enable the White House to prepare a policy statement for the president to take to Congress on May 25, 1961. Galvanized by the Bay of Pigs disaster on top of the Gagarin triumph, and buoyed by the enthusiastic public reaction to Alan Shepard's flight, Kennedy decided to steal the space headlines back from the Soviets by proposing a hugely ambitious objective for NASA and the country.

Why Did the United States Send Men to the Moon in the 1960s? 7

Fig. 1.4. May 25, 1961. President Kennedy announces America's Apollo Moon landing program to a joint session of Congress. Vice President Lyndon Johnson and *(right)* Speaker of the House Sam T. Rayburn listening in were both Texan politicians and were influential in ensuring that NASA's Manned Spacecraft Center was based in Houston Texas. (Illustration courtesy of NASA.)

The speech to Congress (see Fig. 1.4) was long and covered a variety of subjects[6], and it wasn't until near the end that Kennedy called for "this nation to commit itself to achieving the goal, before this decade is out, of landing a man on the moon and returning him safely to earth." Most of the rest of his speech has been forgotten, but that one sentence has been replayed many times over, representing as it does the start of an engineering tour de force that no other country has been able to equal in the 50+ years since.

Despite its brevity, the wording of that commitment was hugely important in defining the technology required. As one senior NASA man explained later – we had only three *sacred* specifications: man, moon, decade [1], p. 176." We will see some examples later where these three words, and especially the word "decade," dictated NASA's technical and management choices in major and sometimes controversial ways.

---

[6] An audio recording of the speech and a transcript are available at https://www.jfklibrary.org/Asset-Viewer/Archives/JFKWHA-032.aspx.

8   Why Did the United States Send Men to the Moon in the 1960s?

Webb now began the task of persuading Congress to approve the funding that would enable NASA to deliver on Kennedy's ambitious commitment.

References

1. Murray, C & Cox, C B, *Apollo: The Race to the Moon*, Simon & Schuster (New York, NY), 1989.

2

The American Rocket

Getting objects and humans into orbit around Earth required huge technical advances. This was first achieved on October 4, 1957, with the launch of *Sputnik-1* by the Soviet Union and by the United States four months later with *Explorer 1*. By the beginning of 1961, 43 spacecraft had been successfully launched – 38 into orbit around Earth and a further five into deep space towards the Moon.

Although the public perception was one of Soviet leadership in this race, the reality was that 34 of the 43 were American, including two of the deep space probes, so both countries had much to be proud of. It was natural therefore that engineers in America and the Soviet Union were eager to tackle the challenge of getting to the Moon (and back) but recognized that the additional technical hurdles would stretch their abilities to the limit.

The key technology was of course a rocket with enough power to carry a large weight into space at sufficient speed to escape from Earth's gravity. There was initially some uncertainty about just how heavy the object taken into space would need to be, but even optimists talked about 100 tons,[1] and pessimists two or three times more. At that time, in and around 1960, the heaviest weight lifted into orbit was about 6½ tons,[2] so an improvement by a factor of 20 or even 50 would be needed.

---

[1] There are at least three different weights that are pronounced "ton." For brevity, I use the word "ton" to signify a weight of 1,000 kg (about 2,205 lbs) instead of "tonne" or "metric ton". Note that in the United States and Canada, "ton" usually means 2,000 lbs, while in the rest of the world it usually means 2,240 lbs.

[2] *Sputnik 7*, launched February 4, 1961, weighed 6.48 tons. It was intended to be a probe to Venus but failed to leave Earth orbit. This remained the heaviest Soviet spacecraft until the Proton launcher became available in 1965, capable of placing 12 tons in orbit. The U. S. Department of Defense SCORE satellite, launched December 18, 1958, weighing 3.96 tons, was the heaviest U. S. spacecraft until NASA's Saturn 1 rocket became available in 1964, capable of placing 17 tons in orbit. *Source: TRW Space Log 1996.*

10 The American Rocket

The basic idea of a rocket is simple enough, similar to firing a canon. Flammable materials are ignited in a chamber open only at one end. The flame produced is made up of hot gases, which rush out through the opening while at the same time push back against the rocket – a bit like a sailor pushing a boat off from the quayside. Standing on the boat you push against the quay wall and the boat is pushed away. The trick is to prevent the flames from melting the engine – that part is called rocket science; in other words it's hard to avoid melting the engine because the flames are hot, white hot. Better materials were gradually developed to cope with this, and pipes containing cool liquids were plumbed around the chamber to cool it down.

Another difficulty with space rockets is that they have to carry so much fuel just to get into orbit that you are lucky if you can carry anything else. For example the Apollo spacecraft sitting on top of the Saturn V rocket on the launch pad at Cape Canaveral made up less than 2 percent of the total weight. So if the Saturn V underperformed by 2 percent nothing would get into orbit. The challenge was to minimize the weight of the rocket's structure including the engines, the fuel tanks and all the various piping, wiring, sensors, latches, etc. One approach that helped was by using the structure itself as the fuel tanks – a so-called monocoque design now commonly used in racing cars – and to thin down the walls wherever you could.

As far back as the 19th century it was realized that you needed to get rid of excess weight as the flight progressed, for example by chucking out fuel tanks when empty. That way the rockets had less dead weight to lift. Most rocket designers went a step further and chucked out the big heavy engines used to get off the ground, keeping only smaller engines to continue the flight. It seemed wasteful to carry an engine that only gets used part of the time, but another factor was that as the flight goes higher and higher the atmosphere becomes more and more tenuous, which changes the way in which the fuel works. Thus the big heavy first stage engines are designed to work in the atmosphere while the smaller second stage engine is designed to work in a vacuum. Some rockets had three or four stages, each smaller than the previous one.

The need to work in a vacuum is of course why a conventional gasoline or diesel or jet engine won't work – they all use the air around them (more precisely the oxygen in the air) to burn the fuel. This would work for the first minute of so of a space mission, but after that you have to provide your own oxygen.

As luck would have it, the fuels that performed best (most thrust for a given weight) tended to be chemicals that were difficult to handle and/or downright dangerous and poisonous. Hydrogen and oxygen mixed together and ignited produce a powerful thrust for their weight, but as gases they take up an enormous volume, so the only practical approach is to cool them down until they become liquids and take up a thousand times less space. Oxygen turns into a liquid when cooled down to −183° C (−297° F) while hydrogen becomes liquid when cooled to −253°C (−423° F). Cooling large volumes of these gases to these very low temperatures (especially the liquid hydrogen one) and keeping them at those temperatures is a mammoth and sophisticated engineering task.

The alternative used in the earliest long-range rockets was a refined form of alcohol or kerosene mixed with liquid oxygen. Pound for pound of fuel neither alcohol nor kerosene is as powerful as liquid hydrogen, but it is a lot easier to deal with. The German V2 rocket of World War II used an alcohol-based fuel combination, while the Soviet rockets that launched Sputnik and Gagarin were kerosene-based. Liquid hydrogen was about twice as efficient by weight as kerosene and was used in the smaller engines of the Saturn V's second and third stages. But the gigantic engines of the Saturn V's first stage were considered too great a leap forward in handling the huge quantities of liquid hydrogen that would be required.

Kerosene was easier to handle since it could be stored at room temperature – no more difficult than storing (at least for a few weeks) the fuel oil needed to heat a house in a basement tank. So this meant that the weight of kerosene carried by the Saturn V was greater than if it had used liquid hydrogen, but this is less critical in the first stage, where in principle you could carry more fuel and thus operate the first stage a bit longer to lift the extra weight of kerosene. There was more reason to avoid kerosene in the second and third stages because the fuel in those stages was dead weight while the first stage was operating. As one of the most famous chroniclers of the *Apollo 11* adventure put it, "It was one thing for kerosene to be obliged to lift its own relatively heavy mass, quite another to have to raise kerosene, which would be doing no work until later [1]."

Rockets burn their fuel in what is essentially a controlled explosion. "Control" is the tricky bit. From 1961 to 1964, engineers at NASA and in industry, especially at Rocketdyne, grappled with how to avoid the Saturn V first stage engine exploding. They had built a prototype that fired up to full power for a few seconds, but left on any longer it destroyed itself. Solving this problem was perhaps the single most important technological achievement that made the Apollo mission possible.

The first stage of the Saturn V rocket was going to be comprised of five F-1 engines (see Figs. 2.1 and 2.2). Each F-1 was to produce 1½ million pounds (680 tons) of thrust, meaning that it could lift that weight.[3] In 1961 this was about six times more powerful than any rocket engine in the United States.

The problem that took four years to resolve was not related to the low temperature of the liquid oxygen nor to the 55,000 horsepower pumps that moved the vast quantities of fuel from the fuel tanks into the rocket chamber, nor even to the materials that made up the chamber and had to withstand the high temperatures.

---

[3] Unless otherwise stated, factual information about the Apollo program comes from [2–6].

12 The American Rocket

Fig. 2.1. First stage of the Saturn V. The nozzles of the five F-1 engines are visible on the left, and the scale is indicated by the humans at the base. (Ilustration courtesy of NASA.)

With a lot of ingenuity and hard work (and money) these problems had all been cracked. The fuel entered the chamber through a flat shower head with some of the pinholes emitting kerosene and others oxygen to ensure a thorough mixing of the two. A pilot light provided a flame that caused the kerosene and oxygen to burn at 5,000° F (2,800° C), and giving rise to a pressure of 1,150 pounds per square inch, or 80 times atmospheric pressure – a ton of kerosene and 2 tons of liquid oxygen per second.

The fuel needed to burn evenly; otherwise pockets of rich and lean fuel would develop, giving rise to pressure and temperature differences around the chamber that quickly bounced off the walls of the chamber, reinforcing each other and getting out of control until the whole assembly disintegrated and exploded. Von Braun and his pioneering team in Germany during the war had encountered this problem and eventually solved it by adjusting the flow of the fuel through the "shower head" (correct name: injector), by adding baffles, by reducing the flow of fuel somewhat and other engineering fixes. After a couple of experimental F-1 engines had been written off at the test range in the Mojave Desert, the team recognized that they couldn't stop the instabilities happening, so they had to accept that they might happen and then stop their effect building up.

Von Braun explained that nobody had yet come up with an adequate understanding of the (instability) process itself, and this "forced the industry to adopt

Fig. 2.2. Wernher von Braun stands beside the base of a Saturn V. (Illustration courtesy of NASA.)

almost a completely empirical approach to injector and combustor development" – a polite way of saying "they used a trial and error approach [7]."

They added baffles into the chamber to damp down the waves of pressure pockets. They tried dozens of different designs of the "shower head" with different-sized holes and different angles of the flow from the holes (see Fig. 2.3). Eventually they got an engine that stayed stable – that is to say, that any unstable performance was damped down in a tenth of a second. To test that this was really stable, they exploded a small bomb inside the rocket chamber while the engine was running to see if the engine could dampen the resulting pressure waves. The bomb would raise the pressure in the chamber suddenly from 1,150 to 4,000 pounds per square inch. And the engine coped with this and kept running. By varying the size of the bombs, test engineers could create instability of different intensities and analyze the ability of the engine to restore stable conditions [8, p. 48, 9].

Fig. 2.3. Amazon founder and CEO Jeff Bezos points out details of the injector plate of an F-1 engine in the Saturn V that carried the Apollo spacecraft into space. The injector was one of many such artefacts rescued in 2013 from the bottom of the Atlantic Ocean by Bezos Expeditions. The first stage of each Saturn V dropped into the ocean when it was ejected by the second stage. The complex arrangement of holes in the nearly 4-foot (110-cm) injector plate and the array of raised baffles (see text) are still visible after the shock of hitting the water and then lying 2½ miles deep for more than 40 years. Jeff Bezos appears again in our story in Chapter 10. (Illustration courtesy of Elston Hill. Used with permission.)

A later NASA report summarized the way the solution was found: The exacting attention to details led to apparently minor changes that actually proved to be of major significance. After careful calculations of the effect, enlarging the diameters of the fuel injection orifices was later judged one of the most important single contributions to improved stability. Other careful changes included readjustment of the angles at which the fuel and oxidizer impinged [7].

The changes to the engine design made it slightly less powerful than it would otherwise have been, but five of the giant F-1s would still be enough to power the first stage of the Saturn V. In early 1965 five F-1 engines were mounted on a test stand and ignited for 6½ seconds. They produced the required 7.5 million pounds of thrust. Observers of the test inside a blockhouse a mile away described their internal organs shaking from the power of the five engines. The noise, too, was fear inducing – it was the loudest sound ever produced by a human short of a nuclear explosion.

The way in which government and industry worked on this problem illustrated the novel management approach pioneered by NASA. It was a sort of halfway house between designing and building everything yourself (the traditional Navy arsenal concept) and contracting things out to industry. It meant that laying blame on one or the other was difficult if problems occurred, but it also meant that both could exchange ideas and information frankly and openly, which certainly saved time. Rocketdyne and NASA staff worked together at the Rocketdyne plant in Los Angeles, California, supported by teams back at von Braun's facility in Huntsville, Alabama, and elsewhere.

The contractor for the first stage of the Saturn V, Boeing, then had the task of assembling five of the F-1 engines plus the massive fuel tanks, pumps, pipes, etc. into a working rocket system. With the F-1 engines themselves in good shape, even though the first stage was an enormous construction, Boeing's task proved to be one of the most trouble-free of the Apollo program.

The contractor for the second stage had a much tougher time. The main problem was that the Boeing first stage contract and the Apollo spacecraft contract were already well underway by the time the second-stage contract was signed. The third stage was going to be an existing rocket and so couldn't easily be changed, and so the second stage requirements needed to be modified numerous times, as the earlier contracts encountered problems and had to change. Typically the Apollo spacecraft would get heavier as its design and construction progressed – adding equipment to deal with unexpected problems. The several tons of extra mass would then be deducted from the permitted mass of the second stage. As noted previously a rocket is mainly fuel encased in a very light structure, so asking for that structure to lose a few tons in weight was seriously difficult to achieve. And the second stage was to be the largest cryogenic rocket ever built – the fuel being liquid hydrogen at $-253°$ C and liquid oxygen at $-183°$ C.

North American Aviation was selected to build the second stage and was forced to find innovative ways to reduce weight. They devised an ingenious honeycomb insulation material that was lighter than metallic alternatives. They also used acid to etch away the aluminum of the tank walls, micron by micron, thinning it to just before the wall would collapse. They combined the oxygen and hydrogen tanks, separating the two only by an insulated metal/plastic barrier, thereby saving four tons. In this way they eventually and with many false turnings delivered a second-stage structure that was less than 10 percent of the weight of the fuel it carried.

The engines in the second stage had been developed by Rocketdyne based on development work begun for the U. S. air force in the mid-1950s – initially intended to be used in rocket-powered aircraft and taken over by NASA when it was created. The first rocket engine using liquid hydrogen and oxygen as fuel that flew in space was that of the Centaur second stage of the Atlas V rocket. Some of the robotic Moon missions that came before Apollo used this rocket, including the Surveyor probes that soft landed on the Moon's surface between 1966 and 1968. The engine for the Saturn V's second stage had to have more than ten times the

16 The American Rocket

Fig. 2.4. Five J2 engines powered the second stage of the Saturn V. Some of the complex plumbing associated with the liquid oxygen and liquid hydrogen is visible. The power of each engine was about a tenth that of an F-1 engine in the first stage, although the nozzle of a J2 is almost as wide as that of an F-1 (see Fig. 2.3). (Illustration courtesy of NASA.)

power of the Centaur engine – 200,000 pounds of thrust as opposed to 15,000 pounds. Although this was still only one seventh of the first stage F-1's 1½ million pounds of thrust, it was a huge increase for what was a very sophisticated engine.

The second-stage engine was called the J-2, and five of them were used just as five F-1s were used in the first stage (see Fig. 2.4). The J-2 not only had to provide the expected level of thrust, it had to be able to stop and then restart in orbit. Being able to restart was anything but simple, especially in the near-vacuum of space, where normal lubricants didn't work (they just boil off) and because of the huge differences in temperature between the super-cold fuel, the room temperature pumps and the red-hot combustion chamber. The F-1 engine in the first stage by comparison could not restart. Once it switched off its working life was over, even if there was fuel left.

NASA's history of the development notes that "At every step of the way, the contractor and the customer [von Braun's team in Huntsville] exchanged information and ideas derived from earlier programs, modifying them for the requirements of the [liquid hydrogen] engine technology, and devising new techniques to implement the design goals of the new rocket powerplant." Rocketdyne had to be especially "encouraged" by NASA to use a technology developed by their competitor, Pratt & Whitney, for the injector ("shower head"). Recall that the injector had been a major stumbling block for the F-1. But eventually Rocketdyne did use the recommended technology, and the problem of burned-out injectors they had been experiencing disappeared [10].

The third stage was confusingly called the S-IVB and had been used on the Saturn 1B rocket since 1965. It used a single J-2 engine with its liquid hydrogen and oxygen fuel combination, with only minor modifications compared to those in the second stage. It, too, had the ability to stop and restart, which would be essential for the Apollo missions. Despite exploding during a test run in 1967 due to inadequate welding, the development of the third stage was straightforward in comparison to the first two stages.

One other unusual aspect of the Saturn V was its control system – called the Instrumentation Unit and built by IBM. It contained three computers checking and adjusting the flight path of the rocket and continually comparing their results with each other. If one of the three gave different results to the other two, it would be ignored. This "two-out-of-three redundancy" was considered a big advance in the use of computers in such a critical role. Von Braun is said to have called the Instrumentation Unit "Saturn's most critical stage" [8], p. 55.

Next let's see how this all went on *Apollo 11*.

References

1. Mailer, N., *A Fire on the Moon*, Pan Books (London), 1970, p. 182.
2. Orloff, R. W., *Apollo by the Numbers*, NASA SP-2000-4029, 2000 (history.nasa.gov/SP-4029.pdf).
3. Anon, *Apollo 11 Press Kit*, July 6, 1969 (www.hq.nasa.gov/alsj/a11/A11_PressKit.pdf).
4. Ezell, L. N., *NASA Historical Data Book, 1958-1968, Vol 2: Programs & Projects*, NASA SP-4012v2, 1988 (history.nasa.gov/SP-4012/cover.html).
5. Van Nimmen, J, Bruno L C, Rosholt R L, *NASA Historical Data Book 1958-1968, Vol 1: NASA Resources*, NASA SP-4012v1, 1976 (history.nasa.gov/SP-4012/cover.html).
6. Various, *The Apollo Spacecraft – A Chronology*, NASA SP-4009, 1969/1973/1978 (history.nasa.gov/SP-4009/cover.htm).
7. Bilstein, R. E., *Stages to Saturn*, NASA, 2004, SP-4206 Part III, chapter 4 (https://history.nasa.gov/SP-4206/ch4.htm).
8. Riley, C., & Dolling, P., *Apollo 11 Owners' Workshop Manual*, Haynes (Yeovil, UK), 2009.
9. Murray, C. & Cox, C. B., *Apollo: the Race to the Moon*, Simon & Schuster (New York, NY), 1989, p. 148.
10. Bilstein, R. E., *Stages to Saturn*, NASA, 2004, SP-4206 Part III, chapter 5 (https://history.nasa.gov/SP-4206/ch5.htm).

3

Apollo 11 – Getting There

Early on the morning of Wednesday July 16, 1969, three of NASA's most capable astronauts enter the elevator at the foot of the gantry to the side of the rocket and ride up the 320 feet (98 m) to the Apollo spacecraft on top of the Saturn V (see Fig. 3.1).[1] Neill Armstrong takes the left seat, Buzz Aldrin the center and Michael Collins is on the right. To squeeze everything in, and to cushion the astronauts during the high-g loads of takeoff and re-entry, the seatbacks are attached to the floor and so the astronauts are lying on their backs staring up. They spend nearly three hours running through checklists, responding to instructions from the controllers, checking dials, and waiting for liftoff.

The Saturn V's fuel tanks are still being filled throughout most of this time. The kerosene being stored at ambient temperature had been loaded up several hours before the astronauts arrived. The liquid hydrogen and oxygen being stored at ultra-low cryogenic temperatures had to wait until the last possible moment to avoid them boiling off. All water vapor and other unwanted gases had to be removed from the fuel tanks ahead of time, using dry nitrogen to clear out the first-stage oxygen tank, and helium to purge the second- and third-stage tanks. Then the liquid oxygen and hydrogen had to be carefully loaded into the pristine tanks – carefully, because they can boil vigorously when entering the relatively warm tank, so the flow of fuel into the tanks has to build up slowly. The fuel had to be continually topped up to replace any that boiled off until just a few minutes before the lift-off.

---

[1] Unless otherwise stated, factual information about the Apollo program comes from [1–5].

Fig. 3.1. Armstrong leads the way to the capsule at Cape Kennedy with an early morning mist still clinging to the ocean far below. (Illustration courtesy of NASA.)

At 9:32 a. m. Florida time the countdown is completed, and the Saturn V's five F-1 engines roar into action (see Fig. 3.2).[2] Onlookers enjoying the 87° F (31° C) weather are kept at a safe 3½-mile (5.5-km) distance and could see the flames belching out at the base, but the sound would not reach them for another 15 seconds. Some say that the rocket wavered as it left the pad, and apparently it is indeed buffeted by the gusty wind. The relatively fragile structure of the rocket

---

[2] The first (central) engine was ignited nine seconds before liftoff, then the others at intervals of a quarter of a second (to lessen the shock), allowing them to build up to full power before the hold-down arms were released and the rocket rose from the pad.

20 Apollo 11 – Getting There

Fig. 3.2. The gantry arms have pulled back as the Saturn V belches flames and leaves the pad. The mini rocket on top of the Saturn V is the Launch Escape System that would have pulled the crew clear if the main rocket had begun breaking up. (Illustration courtesy of NASA.)

was designed to give way to wind rather than fight it. Instead it computes a revised trajectory that takes into account the slightly different position the wind has moved it to. The engines adjust with spurts of acceleration, so the first few seconds are bumpy. "Very rough" said Collins afterwards, "very busy. It was like a woman driving her car down a very narrow alleyway. She keeps jerking the wheel back and forth – a nervous, very nervous lady." Bill Anders on Apollo 8 had described the ride as "being heaved about; like being a rat in the jaws of a giant terrier". After eight seconds the rocket has lifted itself above the gantry tower, and Collins recalled later that "It was nice to know there was no structure around when the thing was going through its little hiccups and jerks [6], p. 198."

For the million or so spectators that are watching the *Apollo 11* launch at Cape Kennedy the 2½ minute burn of the rockets first stage is a spectacular sight, heightened by the ground-shaking sound at the start. By the time the first stage finishes its task it is traveling at close to 6,000 mph (9,600 kmph), more than 40 miles (64 km) high and 70 miles (110 km) away, so you need a good pair of binoculars to see it. The astronauts had been pressed down into their seats by a force four times that of gravity (4 g) generated by the acceleration of the rocket. When the engines cut off after 4 minutes the whole structure of the rocket expands a bit, to which is added the sudden removal of the 4g pushing on their chests plus the firing of small rockets to push the first and second stages apart plus another explosive device cutting the skin of the rocket to allow the two stages to separate smoothly. A second after that the second stage lights up and the acceleration begins again. This whole turbulent sequence "really gets your attention," as *Apollo 16* astronaut Ken Mattingly put it. "Man, that was quite a sequence" was how *Apollo 10*'s Tom Stafford described it. "I thought I was going through the instrument panel," said Fred Haise while on *Apollo 13*.

The acceleration of the second stage exerts a hardly noticeable 1 to 1½ g force on the astronauts, and this continues until nearly ten minutes have gone since launch. *Apollo 11* is now 120 miles (190 km) high, traveling at 14,600 mph (23,500 kmph) and 1,160 miles (1,860 km) out over the Atlantic Ocean. Even with a powerful telescope the onlookers at the Cape can see nothing. The dramatic yellow (dirty) kerosene flames of the huge first-stage engines (see Fig. 3.3) are absent from the second and third stages because their hydrogen fuel burns with a clean flame that is almost completely invisible (see Fig. 3.4). The acceleration of the second stage may be relatively modest, but on some of the Apollo flights (not to any great extent it seems on *Apollo 11*) there were other features of the ride to "get their attention." *Apollo 8* and *Apollo 13* both experienced an up-and-down shaking called the "pogo" effect because of its similarity to movement on a pogo stick. Pogo in the first stage had almost torn the unmanned *Apollo 6* test flight apart in 1968, so changes were made to eliminate it throughout the rocket, but never with complete success.

One of the more enjoyable aspects of the second stage burn is that the astronauts finally could see out of all of the windows. They had been sitting in their seats in the Apollo Command Module, backs to the floor, looking out the windows, some of which showed a blank surface – the inside of the protective cover that shielded the Apollo capsule from the heat of air friction during the first few minutes. Above the cover was the needle-shaped Launch Escape System rocket (see Fig. 3.2) that, in the event of a problem with the first stage, would fire for eight seconds and propel the whole Apollo capsule at a back-stiffening 7 g far enough away from the main rocket to allow it to float down to Earth on its parachutes.

Fig. 3.3. The rocket seems to ride the yellow flame of the kerosene fuel in this photo from an air force plane. (Illustration courtesy of NASA.)

Once the second stage burn has begun, a small rocket at the top of the Launch Escape System pulls it and the protective cover clear of the Saturn V, unblocking the view through all of the windows. Armstrong confirms that this is so with his matter of fact "Houston, be advised the visual is Go today," while Collins gives a bit more insight into the changed perspective with "Yeah, they finally gave me a window to look out." Five minutes later he takes advantage of his new-found view to comment on the weather below: "Well, it looks like a nice day for it. These thunderstorms down range is about all."

After its job is done the second stage is jettisoned with little of the turbulence when the first stage was dropped, and the third stage fires up. What remains of the vehicle is just one-sixteenth of the 3,000 tons[3] that lifted off, so the single J-2

---

[3] There are at least three different weights that are referred to as "ton." For brevity, we use the word "ton" to signify a weight of 1,000 kg (about 2,205 lbs) instead of "tonne" or "metric ton." Note that in the United States and Canada, "ton" usually means 2,000 lbs, while in the rest of the world it usually means 2,240 lbs.

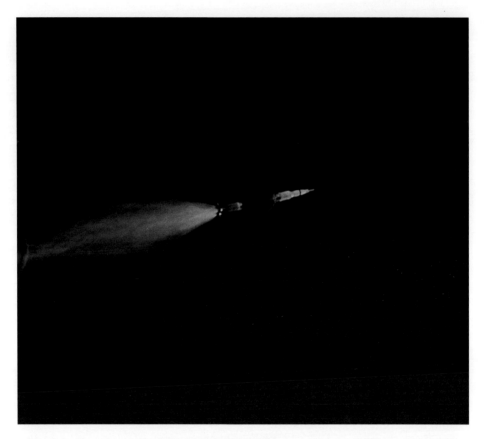

Fig. 3.4. Forty-two miles high, the hydrogen of the second stage lights up on the right, emitting a colorless flame (you can just see its faint blue shock wave) while the first stage falls away on the left, still spilling out leftover propellant. (Illustration courtesy of NASA.)

engine in the third stage is sufficient to accelerate it horizontally for 2½ minutes at a gentle ½ g or so up to the 17,500 mph (28,000 kph) needed to go into orbit about 120 miles (190 km)high[4].

On the *Apollo 11* mission Neil Armstrong reported that the "Saturn gave us a magnificent ride. We have no complaints with any of the three stages on that ride. It was beautiful." He then confirmed to Mission Control that there had been no turbulence at the end of each stage (more precisely: "no transients at staging of any significance").

---

[4] This velocity is a so-called space-fixed velocity since the spacecraft is now in orbit. Velocity figures for earlier in the flight were so-called Earth-fixed, i.e. relative to being stationary on the launch pad.

The Apollo capsule still attached to the Saturn V third stage orbits Earth for about two hours, completing not quite two circuits of the globe. Their time is filled with running checks on the equipment and updating the computer with information radioed up from Mission Control. The events are made more complicated by there being only intermittent communication between the ground and the capsule, with losses of contact over wide stretches of ocean. In due course as they come north and east over the Pacific Ocean, heading in the direction of the west coast of the United States for the second time, the third stage re-ignites its J-2 engine and takes them out towards the Moon. The crew notices the effect of the engine because they go from being in zero-g, where everything floats around unless tied down, to a gentle ½ g for about six minutes. That is sufficient for the combined third stage plus the Apollo vehicle to reach a velocity of nearly 25,000 mph (40,000 kmph). This isn't quite fast enough to escape Earth's gravity completely and head for the stars, although the third stage could have raised the velocity enough to achieve that. The new velocity will take Apollo past the Moon, and if they fail to stop there, Earth's gravity will in fact pull it back home (as happened with *Apollo 13* the next year).

Aldrin, Armstrong and Collins become only the seventh, eighth and ninth people to travel this path, and Collins for one acknowledges the significance of heading away from Earth into the gravitational embrace of another body. He recorded in his autobiography that he felt there should have been some more momentous announcement during the *Apollo 11* flight than the mundane exchange that actually took place:

Mission Control: "You are Go for TLI [trans lunar injection]"
Collins in Apollo 11: "Thank you."

It takes three days to reach the Moon, and it starts out with a busy schedule for the astronauts. They are still attached to the third stage of the Saturn V, and they have not only to detach from that but also to grab hold of the Lunar Module from the protective cocoon-like shell in which it has been shielded from the turbulence of its journey through the atmosphere and the violence of the rocket blasts and ejections.

Let's just review the various bits of hardware that have made it this far. The first and second stages of the Saturn V have been jettisoned and fallen back into the ocean, leaving the third stage, above which is the Lunar Module in its cocoon. Above that is the Service Module, and finally at the top, is the Command Module, where the astronauts sit. Now that the third stage has done its job of sending them towards the Moon, the idea is to get rid of its dead weight and for the Command, Service and Lunar Modules to continue as a single vehicle. The Command and Service Modules have been locked together tightly from the start, but the Lunar Module has to be extracted from its shell. The astronauts unhook their craft from the rocket, turn their vehicle around and inch back down to the rocket in order to dock with the Lunar Module. Once they hear the solid "click" of the docking mechanism they inch away from the rocket, taking the Lunar Module with them (see Fig. 3.5).

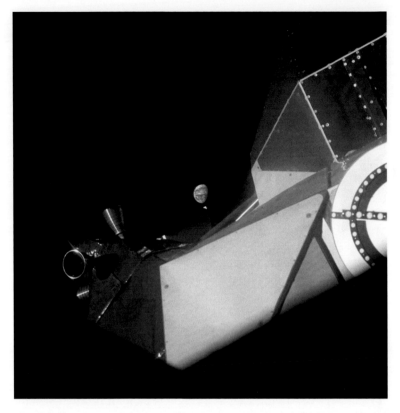

Fig. 3.5. Heading for the Moon, looking back at Earth past the Lunar Module now attached to the main Apollo spacecraft. (Illustration courtesy of NASA.)

The inconvenience of keeping the Lunar Module in a separate protective structure was dictated by the fact that for every pound of material that is landed on the Moon's surface, 500 pounds have to be added to the mass of the launcher – extra fuel in the first, second and third stages, and in the Lunar Module itself as it slows down and lands on the Moon's surface. By shielding the Lunar Module from the violence of the launch, its designers could make it more fragile and thus lighter [6], p. 315.

It may seem surprising that given the enormous speed of Apollo as it left Earth (almost 25,000 mph) it should take it three days to reach the Moon. Although it sets off for the Moon at a fast clip, Earth's gravity is constantly (although decreasingly) pulling on the craft and thus slowing it down. Two and a half days later they are within 35,000 miles (56,000 km) of the Moon, at which point Earth's gravity has decreased and the Moon's gravity has increased (as they approach it), so that the two gravity pulls are equal. The spacecraft speed is now down to less than a tenth of its starting value, although still about 2,000 mph (3,200 kmph). From then on the Moon exerts a greater pull on the craft than Earth does, and their speed starts to increase again.

As they approach the Moon, Armstrong sees what he later said was the most impressive sight of the mission – flying through the shadow of the Moon, that is to say, having the Moon cross in front of the Sun. The Moon was by this time just 12,000 miles (19,000 km) away and loomed very large in their sight. The Sun passing behind its edge produced a halo effect, which the crew describe as eerie. The surface of the Moon is now lit only by earthshine (the light from Earth), and this is sufficient for them to make out craters on the Moon's surface.

Furthermore, while the Sun is behind the Moon they are able to see the stars clearly for the first time. Previously the Sun's glare reflected off the spacecraft was dazzling, making it difficult to see stars or to make out details on the Moon.

Fourteen hours later they reach the Moon and are pulled by its gravity into a tight arc around it. They lose sight of, and contact with, Earth as they swing behind the Moon, traveling now at over 5,000 mph (8,000 kmph).

Command and Service Modules (CSM)

Let's take a closer look at the vehicle they are in. For most of the journey the Command and Service Modules act as a single vehicle. It is only when they return to Earth and prepare to enter the atmosphere at 25,000 mph (40,000 kmph) that they jettison the Service Module. Thus only in the last hour or so of their week-long journey to the Moon and back do the astronauts rely solely on the resources of the Command Module. Let's therefore consider these two modules as one – dubbed the CSM (see Fig. 3.6).

The astronauts live, sleep, eat, cook, work and travel in the CSM – it's a miniature home, office and car all in one. But more than that it supplies them with breathable air (supplying oxygen, removing carbon dioxide), with food and water, with waste extraction and garbage removal, with fuel for their vehicle, with radio and TV, etc. So it's a whole world in itself – for a week.

To complicate matters for its builders, the CSM had to be made as lightweight as possible so that it could be accelerated to 25,000 mph. To make sure it got the best, NASA evaluated the designs of several companies who had put together proposals to build the CSM. The NASA evaluation panel recommended the proposal made by the Martin Company, but the NASA administrator, James Webb, and two of his deputies decided to give the contract to North American Aviation, which had come second in the evaluation. Three months after that North American

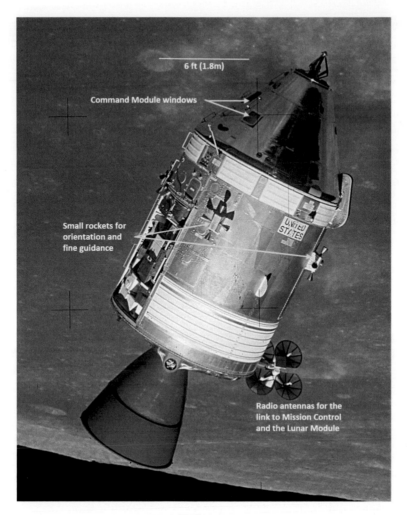

Fig. 3.6. The Command and Service Module as seen by the Lunar Module (*Apollo 15*, July 1971) above the Moon. The Command Module is the gold-tinted cone at the top; the Service Module is everything else with the rocket nozzle at the base. The approximate scale is shown at the top. (Illustration courtesy of NASA/author.)

Aviation gave a vending machine contract to a company called Serv-U, one of whose owners, Bobby Baker, was a protégé of Senator Robert S Kerr, who in turn was a former boss of the NASA administrator.

The critics said that Webb's decision to choose North American Aviation was because of his ties to Senator Kerr, and this became the subject of what became known as the Bobby Baker investigation by the Senate Rules Committee in 1963.

Certainly the contractor made hard work of the development. This came to a head with the tragic events of January 27, 1967, when the *Apollo 1* capsule was being tested on top of a Saturn 1B rocket at Cape Kennedy a month before it was due to be launched. A fire broke out inside the capsule, killing the three astronauts inside – Gus Grissom, Roger Chaffee and Edward White. The accident investigators found a litany of errors by North American Aviation, including bare electrical wires that lay side by side and other examples of shoddy and dangerous practices. In addition to poor workmanship, the design of the CSM was unsafe, for example making it impossible for the astronauts to get out in time. NASA, too, bore some blame for design faults, since they had to approve the design before it was built.

North American Aviation were kept on as contractor, but they had to make many changes to the design of the CSM, and just as importantly to improve the management of the work to ensure that the final CSM was effective and safe. Let's look a little more closely at what they produced.

Service Module

The Service Module had a large rocket engine (and several small ones), fuel tanks, oxygen tanks, water tanks, electricity generators (hydrogen fuel cells) and radio and electronic equipment. It was the engine room and store for the Command Module.

The Service Module engine was considerably less powerful than those of the Saturn V but was still able to generate almost 10 tons of thrust. Its design was intended to reduce the possibility of failure in several ways. First and foremost the fuels used did not need to be ignited. As soon as the propellant (dimethyl hydrazine) came into contact with the oxidizer (nitrogen tetroxide) the mixture exploded. These chemicals are not household names primarily because they are very corrosive and toxic, and therefore dangerous. A second simplification was to eliminate pumps to drive the fuel into the engine – the moving parts in a pump were always a potential source of problems. In the zero-g of space the fuels would not "fall" into the engine; instead they had to be pushed. This was done by having helium gas in the fuel tanks under pressure. The helium kept a steady pressure on a flexible membrane, which in turn pushed the fuel out of the tanks. When sitting on top of the Saturn V at Cape Canaveral about a third of the Service Module's weight was fuel – 16 tons of it.

The Service Module generated electricity for itself and the Command Module using hydrogen fuel cells. The earlier Mercury and Gemini flights had used batteries, but the relatively long duration of the Apollo mission ruled out batteries. Fuel cells were the choice because not only did each of the three fuel cells produce one kilowatt of power, as a byproduct they produced water used by the crew for drinking and washing (hydrogen mixes with oxygen while passing over a catalyst, resulting in electricity and water). About 50 gallons of water were produced this way during a mission.

Astronauts need air to breathe – or at least the oxygen in air. On the launch pad the air was an oxygen-rich version of the Florida atmosphere – 60/40 oxygen / nitrogen instead of the normal 21/79. As the rocket took them into orbit, the nitrogen was gradually removed, leaving pure oxygen at a third of normal atmospheric pressure, which was enough to keep the crew's blood oxygen levels normal. The astronauts had to start breathing pure oxygen three hours before launch to flush nitrogen out of their blood and thus avoid the bends during the reduction in air pressure (decompression). The low atmospheric pressure inside the Command Module allowed walls to be thinner and thus lighter.

The environmental control system in the Service Module kept the air pressure in the command module steady and also performed some magic by passing it through canisters of lithium hydroxide to scrub it of carbon dioxide. It was then pumped back into the cabin to be re-breathed. Whereas a scuba diver uses a tank of air every hour, the equivalent amount of oxygen lasted 15 hours in the Apollo capsule [7]. The Service Module itself was unpressurized (to save weight) – the air inside it gradually leaked into space, so its equipment had to function in a vacuum.

Controlling the temperature was another headache. In sunlight, temperatures reached a searing 390° F (200° C, double the boiling point of water) while in the shade it could be an icy –240° F (–150°C). Thus one side of the spacecraft would be very hot, the other side very cold, and meanwhile the inside of both Command and Service Modules had to be kept at about room temperature. To save weight, rather than add special reflective and insulating coatings to the outside, the spacecraft slowly rotated all the way to the Moon and back, thereby avoiding the buildup of heat or cold. The maneuver was known as the barbecue roll, which ensured that the crew and the equipment were "cooked" just right throughout the flight. An added bonus for the crew was that the view out of the windows kept changing gradually.

The radio and electronic equipment in the Service Module was a critical but relatively unsung part of the mission. There was an inertial measurement unit that was a chunky and clunky version of the miniature device in your smartphone that knows which way is up and rotates your screen accordingly. There were small telescopes with cameras attached that detected designated stars, the Sun, the Moon and Earth's horizon. With this information the guidance computer could work out which way the craft was pointing and adjust the inertial measurement unit if needed. A smartphone or tablet app such as SkyView can identify individual stars in the night sky using much the same technique.

The guidance system was based on what had been developed for the Navy's Polaris submarine launched missile, under the leadership of the Massachusetts Institute of Technology (MIT) and the legendary Professor Charles Stark Draper, who had conceived the key technical features over the previous 30 years. The Apollo version was much more complicated than that of Polaris (whose flight lasted less than 20 minutes and remained close to Earth) but benefited from computers that were getting more powerful each year – although still minuscule compared to today's.

As well as keeping the spacecraft pointing in the right direction, in principle the guidance computer could steer the spacecraft based on these telescope and inertial measurements, but in practice that was a back-up way of navigating that was never used – fortunately since it was not very accurate.[5] The key pieces of equipment for navigation were the radar transponder that enabled the radar stations on Earth to track the CSM and the radio receiver that fed that information from the ground to the crew.

We think of radar as measuring how long it takes for a radar station on ground to pick up the reflection off a vehicle of a signal it has just transmitted. In a hostile military context the echo is the signal that bounces off the skin of the enemy's vehicle. That is also how a speed trap radar works on the highway – the body of your car reflects back the signal from the radar gun held by a police officer or in an overhead gantry. However if the vehicle is not your enemy you can get a much stronger signal if the vehicle picks up your signal and transmits it back at full strength instead of just reflecting it. The returning signal will be much stronger, allowing you to track a vehicle at a much greater distance. This technique is routinely (indeed mandatorily) used in civil aviation, ensuring that air traffic controllers have reliable information on the location, speed and direction of travel of the aircraft they are managing. The same is true for ships at sea, especially near harbors. Space "ships," too, cooperate with the ground radar by re-transmitting the radar signal and thus can be tracked for thousands, even millions, of miles.

The equipment on the vehicle that transmits back the radar signal to its source is called a transponder – transmitter/responder. So Mission Control in Houston would process all that radar data in their much more powerful computers and send details to the Command Module which the astronauts would enter into the guidance computer.

---

[5] I led a small team at TRW Systems in Houston that analysed the accuracy of these measurements after the Apollo, 8, 10, 11 and 12 missions. Our results confirmed the decision not to navigate using the sensors and computer in the CSM.

Command Module

It was cramped for the crew inside the Command Module, and the food was nothing to write home about, either. They all lost weight during the trip: Armstrong 9 lbs (4 kg), Collins 8 lbs (3.5 kg) and Aldrin 6 lbs (2.8 kg). The sleeping accommodation is quickly described – there was none. Oh, and there was no bathroom.

What it did have was three seats, five windows, a hatch, a complicated dashboard (see Fig. 3.7) plus a docking mechanism that could be removed to allow Armstrong and Aldrin to crawl into the attached Lunar Module – a bit like crawling through the trunk of your car into a tube to get to an attached caravan at the back of a car while it's traveling at breakneck speed.[6]

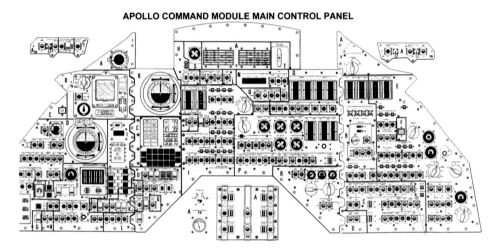

Fig. 3.7. Buttons, dials, switches, levers, lights and small circular displays make for a mind-bogglingly complex dashboard to be operated by the crew. Overall flight control was on the left hand side, main engine and electrical power on the right and air conditioning and fine guidance in the center. (Illustration courtesy of NASA.)

The Command Module was in the shape of a cone; the back, almost flat, part of the cone was reinforced to withstand the 25,000 mph (40,000 kmph) re-entry into Earth's atmosphere. On the launch pad, while still subject to the effect of Earth's gravity, the three crew members were lying on their backs while sitting down

---

[6] You can see for yourself by visiting one of the sixteen Command Modules that returned from space and are located at museums around the US and abroad. See https://nssdc.gsfc.nasa.gov/planetary/lunar/apolloloc.html for details.

staring straight ahead (upwards) at the tip of the cone. Once someone was in space and in zero-g, up and down had no meaning, so any feeling of sitting on the floor soon vanished.

The space not taken up by equipment was roughly 210 cubic feet (6 m^3) in size, more than the inside of a large car although shaped as a cone about 12 feet across and 6 feet high (3.6 m by 1.8 m).

A few pages ago we left the astronauts as they disappeared around the back of the Moon, out of sight of Earth. Their path takes them to within 92 miles (148 km) of the Moon's surface, which is where they fire up the engine for six minutes to reduce their speed by 2,000 mph (3,200 kmph), using up almost 11 tons of fuel (reducing the CSM's weight by 40 percent). If the engine fails to fire the worst that can happen is that they whip around the Moon and head straight back to Earth. The crew know that the engine worked by the kick in their backs as it lights up, and they can tell again when it stops after the required six minutes by the fact that they return to being weightless. Back in Houston, Mission Control has to wait nervously for the CSM to emerge from behind the Moon before they find out if the engine has done its job.

They are now orbiting the Moon, and on the third pass around the back of the Moon they fire the engine again for 17 seconds to place the CSM in an orbit that as nearly as possible duplicates that of *Apollo 10* two months earlier. This means they are about 65 miles (105 km) high as they whiz around, the craters and mountains below slipping past the window at a perceptible and steady pace.

They go around a total of thirteen times, making the preparations for the landing. *Apollo 10* had trail-blazed in May and found that the Moon's gravity had dragged them off their planned course by several miles. By following the same path as *Apollo 10,* the *Apollo 11* crew expected that their path would repeat that of *Apollo 10* and thus be predictable.

Each of the thirteen orbits around the Moon takes about two hours, and so it is a full day before the Lunar Module is ready to separate from the CSM and head for the surface. Michael Collins remains alone in the CSM in orbit 100 miles (160 km) high while Armstrong and Aldrin fly down.

How to Land on the Moon (and Return)

The early science fiction stories of travel to the Moon envisioned a rocket flying direct from Earth to the Moon and back (see Fig. 3.8). Why then the complicated arrangement with a CSM separate from a Lunar Module? The answer came down to money. A direct trip to the Moon would require an enormous rocket leaving from Earth, two or more times the size of the Saturn V. It wasn't clear that such a beast could be built, given that the Saturn V was already a huge stretch beyond

Fig. 3.8. A hundred years before Apollo (1865) French novelist Jules Verne imagined how humans could reach the Moon. He proposed a very large Earth-bound gun that fired a projectile-style capsule and its crew to the Moon. Although he recognized that Verne's idea was unworkable, Russian scientist Konstantin Tsiolkovsky was inspired by the novel to analyze the science of spaceflight and rocketry thoroughly, publishing his groundbreaking results in 1903

what was previously available, so a direct trip launcher would cost unknown sums of money and take a long time to develop.

The early Apollo designers looked at how best to make the trip. The initial consensus was that if you couldn't build a big enough rocket, then use several Saturn V rockets and assemble the Moon craft in space. Two or perhaps three Saturn V launches would be needed. Von Braun was one of those who favored this approach, not least because it was consistent with his vision of building space stations in Earth orbit to be used as staging posts to exploring the planets. Assembling a lunar landing spacecraft in Earth orbit would be an early demonstration of what he expected to be a much repeated general approach to reaching Mars and beyond.

Not everyone agreed with this approach. Some were against it simply because it would be an expensive, risky and long program. Assembling large, hazardous, complex, precision machines in orbit was a completely untested idea – pure science fiction in the view of critics. Even the optimists admitted that there was little chance of completing this proposed approach by 1970.

The alternative that emerged sounded initially even more exotic and almost bizarre: direct launch into orbit around the Moon, leave the main CSM craft orbiting the Moon and send a small craft down to the surface, which returns to the CSM in orbit around the Moon which then returns to Earth. You avoided landing a heavy spacecraft on the surface that would have to bring enough fuel with it to propel itself off the surface and out of the Moon's gravity to get back to Earth.

This discussion was taking place in 1960-62, before space vehicles had rendezvoused in Earth orbit (that wouldn't happen until NASA's *Gemini 6* and *Gemini 7* double mission in 1965), so imagining that they could do so in orbit about the distant Moon seemed fanciful – "ambitious" might be a polite way of phrasing it.

However the numbers seemed to add up. The Apollo mission could be achieved with a single Saturn V launch instead of two or more. In broad terms this halved the cost of the program! And apart from the computational challenge of rendezvousing in orbit around the Moon, it didn't need any radically new technology – so it could perhaps fit with the 1970 deadline.

Eventually von Braun was convinced that rendezvous in Moon orbit was the right answer, and that settled it. A program with just a single Saturn V launch became the target, and the Lunar Module was born. In November 1962, just four months after this decision, Grumman Engineering Corporation was selected as prime-contractor for the Lunar Module.

The decision to go with a single Saturn V launch led to the concept of the Lunar Module as a separate spacecraft (see Fig. 3.9). The spindly legged vehicle was not elegant or comfortable. Two astronauts would spend a day in it (rising to more than three days for the final missions, *Apollo 15-17*), so fittings were kept to a minimum. There were no seats. They stood to operate the vehicle and lay as best they could to sleep. But technically the Lunar Module was highly sophisticated.

The rocket engine used during descent to the surface was especially advanced. Normally a rocket engine is either on or off, but in this case you had to be able to vary the power of the engine so that the Lunar Module could descend faster or slower, and even hover.[7]

---

[7] Declaration of interest: the engine was designed and manufactured by TRW, my employer at the time.

Apollo 11 – **Getting There** 35

Fig. 3.9. The spindly legged part of the Lunar Module remained on the Moon's surface while the crew cabin on top had its own rocket engine to get it back to the CSM in orbit above. (Illustration courtesy of NASA.)

In fact, the Lunar Module was actually two vehicles – the descent stage and the ascent stage, each with its own separate rocket engine. Thus the part that took off from the Moon was absolutely the minimum weight to get back into orbit around the Moon and meet up with the CSM, leaving behind the legs, engine and fuel tanks for the landing, the altimeter that told them how close they were to the ground, TV cameras and communication equipment, their backpacks, garbage and so on.

First, the astronauts had to get down safely from orbiting the Moon in the CSM to reach the surface, softly, in the Lunar Module.

Having crawled into the Lunar Module, Armstrong and Aldrin cast themselves adrift from the CSM and fired up the descent engine. As they pulled away from Collins in the CSM, Armstrong's laconic drawl rang out with "the Eagle has wings," referring to the name the crew had christened the Lunar Module. Collins replied that they were flying upside down, but not to be outdone Armstrong came back with "Well, someone is upside down."

References

1. Orloff, R. W., *Apollo by the Numbers*, NASA SP-2000-4029, 2000 (history.nasa.gov/SP-4029.pdf).
2. Anon, *Apollo 11 Press Kit*, July 6, 1969 (www.hq.nasa.gov/alsj/a11/A11_PressKit.pdf).
3. Ezell L N, *NASA Historical Data Book, 1958-1968, Vol 2: Programs & Projects*, NASA SP-4012v2, 1988 (history.nasa.gov/SP-4012/cover.html).
4. Van Nimmen, J., Bruno L C, Rosholt R L, *NASA Historical Data Book 1958-1968, Vol 1: NASA Resources*, NASA SP-4012v1, 1976 (history.nasa.gov/SP-4012/cover.html).
5. Various, *The Apollo Spacecraft – A Chronology*, NASA SP-4009, 1969/1973/1978 (history.nasa.gov/SP-4009/cover.htm).
6. Maile, N., *A Fire on the Moon*, Pan Books (London), 1970.
7. Riley, C. Dolling, P., *Apollo 11 Owners' Workshop Manual*, Haynes (Yeovil, UK), 2009, p. 83.

4

The Eagle's Journey

An eagle was chosen as the mission emblem, as shown in the *Apollo 11* badge or patch (see Fig. 4.1), highlighting that it was American. The patch was also unusual in omitting the names of the astronauts (the only Apollo mission patch to do so), apparently at the suggestion of Collins, who wanted the credit to be shared with the thousands of people whose work had made it a success.

Fig. 4.1. The American eagle symbolized the *Apollo 11* mission and also was used as the name of the Lunar Module. (Illustration courtesy of NASA.)

It took about half an orbit to slow the Lunar Module down from 3,750 to about 410 miles per hour (6,035 660 kmph), dropping down all the time to 10,000 feet (3,000 m).[1]

Armstrong at the controls couldn't see the ground because the base of the Lunar Module was facing in the direction they were traveling – initially at 3,750 miles per hours. It wasn't until they came down to within 10,000 feet of the ground and the forward speed had dropped to 400 miles per hour that the Lunar Module started to tip upright while still descending (see Fig. 4.2). By about 500 feet they were vertical with the feet pointing down, and they were then descending more or less vertically, and slowly (see Fig. 4.3). Armstrong and Aldrin realized that they didn't recognize the craters they were passing, and this was both a surprise and a worry. The *Apollo 8* and *10* astronauts had brought back high-quality close-up photos and videos of the expected route to their landing site, and it had dozens of easily recognized craters and groups of craters that they had memorized.

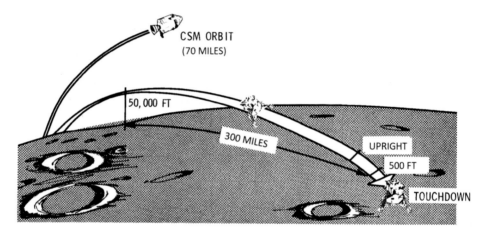

Fig. 4.2. As they emerge from behind the Moon the Lunar Module has dropped to about 10,000 feet (3,000 m) when it is overtaken by the CSM above. The Lunar Module's feet have been facing forward to allow the engine to slow it down, but at this point it starts to rotate to an upright position, which is achieved when about 500 feet (150 m) high, giving the pilot a view of the ground below. (Illustration courtesy of NASA.)

Back in the day, Johannes Kepler and Isaac Newton had worked out that you would orbit around the Moon or a planet in an ellipse – an elongated circle. That prediction would have been accurate if the Moon had been a perfect sphere with no lumps or bumps, but even from Earth you could tell that the Moon was very

---

[1] Unless otherwise stated, factual information about the Apollo program comes from: [1–5].

lumpy and bumpy, and the gravitational effects of these irregularities would change the orbit from an ellipse into an ellipse-with-wiggles.

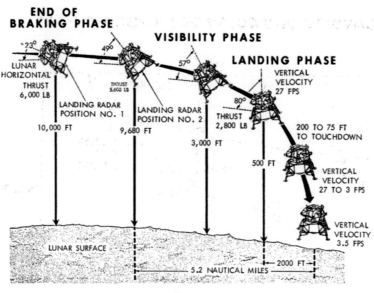

NOMINAL DESCENT TRAJECTORY

Fig. 4.3. This shows how the descent SHOULD have gone – straight down from about 500 feet high. Armstrong instead had to fly the Lunar Module like a helicopter to find a flat, boulder-free area. (Illustration courtesy of NASA.)

When NASA's five Lunar Orbiter robotic satellites went into orbit around the Moon in 1966 and 1967 the pictures they sent back confirmed that the Moon was anything but smooth, and sure enough their path around the Moon was tweaked by those lumps and bumps. The tweaking was most pronounced when they got close to the Moon's surface, which the technicians call "perilune," so the tweaking was nicknamed "perilune wiggle." By giving it a childish name, the engineers and scientists trying to explain it made it almost endearing, but it was in fact a big cause for concern.

Apollo 8 and then *Apollo 10* flew as close as they could over the path that *Apollo 11* would take so that they could predict how *Apollo 11*'s orbit would wiggle. However, all three missions experienced different changes to their orbits presumably because their paths around the Moon were not absolutely identical, and even a small difference in the orbit changed the gravitational "tweaks" significantly.

The consequences were as follows. The CSM does thirteen orbits of the Moon before the Lunar Module leaves it and heads for the surface. On orbit number twelve Mission Control gives the CSM instructions to fire the engine and adjust the orbit so that on the *next* orbit they will be in the right position for the Lunar Module to descend. The seemingly simple task therefore is for Mission Control to be able to predict the CSM's path one orbit (two hours) ahead. But perilune wiggle made that impossible – at the time. The result is that when the Lunar Module "gets down among the craters," as *Apollo 10*'s Gene Cernan had put it, they didn't recognize the landscape because they were several miles off course. Cernan had described being "among the boulders" (although actually 10 miles above them), and *Apollo 11* finds itself in an unsuspected boulder field that Armstrong has to fly out of to find a relatively flat area to land – and uses up most of his spare fuel in doing so!

It turns out that Armstrong and Aldrin were 2 miles south and 4 miles west of their intended landing site. The irregular masses beneath the Moon's surface had shown once again their power to tweak the trajectory of a low-flying spacecraft.

It had taken NASA a couple of years to verify that perilune wiggle was a gravitational effect and not something else. Other possible causes had been considered when the wiggle was first noticed in 1967, from the mundane to the bizarre. Perhaps it was just some glitch in the radar data due to being close to the Moon's surface and echoing off it. Or did the Moon have some inherent property that garbled radar data? Or was there some effect of Einstein's general relativity that was not being accounted for?

The answer was convincingly presented by two scientists at the Jet Propulsion Laboratory near Los Angeles in 1968, Frenchman Paul Muller and Californian Bill Sjogren. They plotted the occurrences of the irregularities (wiggles) in the measured velocity of the lunar orbiter satellites on a map of the Moon, and lo and behold the biggest irregularities were found over the big circular features, the so-called seas ("mares" in Latin). The two biggest circular features form the eyes of the Man in the Moon (as seen in northern latitudes) – the seas of Serenity and Showers. Others were the seas of Crisis, Nectar and Moisture, and just peeking around the eastern edge of the Moon, the Eastern Sea.[2] Their circular form meant that they had been hit by a large body from outer space whose mass was buried deep beneath and which caused heavy material from deep inside the Moon to rise towards the surface. These impacts had taken place 4 billion years ago, and unlike Earth, the Moon had no tectonic activity that could mix them in with the main mass of the Moon. The multitude of craters on the Moon's surface were also circular and also the result of infalling bodies from outer space, but they lacked the huge scale of the seas and thus did not release the deep heavier material, and their individual gravitational effect was not detectable.

---

[2] Their Latin mare names are Serenitatis, Imbrium, Crisium, Nectaris, Humorum and Orientale, respectively.

The maps produced by Muller and Sjogren showed that circular mass concentrations ("mascons," as they became known) caused the wiggles, but the maps weren't in a form that could be used to program a computer. It would be 20 years before data from more satellites orbiting the Moon and computers powerful enough to analyze their data would be able to provide a mathematical equation that both captured the detail of the gravity variations and was in a form that a computer could use [6]. Back in 1969 the computer in the Lunar Module was tiny by modern standards and could only accommodate a very simple formula for the Moon's gravity. The computers in Mission Control were better but lacked an adequate formula of the gravity field.[3]

Inside the Lunar Module the first part of Armstrong and Aldrin's final descent to the Moon's surface had been flown "blind." They were at 50,000 feet (15,000 m) traveling at more than 3,000 miles per hour (4,800 kmph) when they put "the brakes on" – that is to say, they pointed the engine forwards and ignited it. They were standing inside the Lunar Module (no seats, remember) on top of the engine, so with the engine pointing in the direction of the orbit, they were tilted at 90 degrees to the Moon's surface.

Initially they were face down to the Moon's surface and noticed with some trepidation that one of the craters they were expecting to see, Maskelyne W, had come into view more than two seconds later than expected, which meant that they were at least 2 miles off course (3,600 mph is one mile per second). Then they flipped over, engine still facing forward but now face out into space, even catching a glimpse of Earth a quarter of a million miles away. This orientation meant that as they turned upright at the end of the braking they would be facing forward and could see the landing site (see Figs. 4.2 and 4.3). Meanwhile the landing radar on the Lunar Module was pointing down and could tell them how far below the surface was.

The braking continued for 250 miles (400 km), at which point they were 13,000 feet (4,000 m) up and traveling at about 600 miles per hour (1,000 kmph). Now they flipped the Lunar Module upright and began to descend. At 7,000 feet (2,100 m) they were moving forward at about 50 miles per hour (80 kmph) and could see where they were going.

After the mission Armstrong said that they didn't actually take much notice of the landing area until they were down to 2,000 feet (600 m) because their computer was giving off a series of alarms. One of the alarms in particular (entitled "1202" – pronounced as twelve-oh-two) was one they hadn't been exposed to in the many pre-flight simulations. And most of the folks at Mission Control were also puzzled.

---

[3] My team at TRW Systems in Houston was one of several that spent much of 1968 trying out various mathematical models for the Moon's gravity and various durations of radar data to try to find a mix that accurately predicted the CSM orbit two hours ahead, but without success.

The Controller keeping an eye on the Lunar Module computer was Stephen G. Bales, and he did recognize it as signaling that the computer was overloaded – the software was trying to do too many things at once. He knew that if the alarm didn't occur too often the computer would prioritize its tasks and do those necessary for the landing while postponing other less essential tasks. The first 1202 alarm came soon after the braking maneuver began and took Mission Control an agonizing 30 seconds to respond with "We're GO on that alarm," meaning the crew could ignore it.

Fig. 4.4. Buzz Aldrin inside the Lunar Module before separating from the CSM. Years later Aldrin seems to have admitted to unwittingly causing the computer alarm signals that nearly aborted the landing (see text). (Illustration courtesy of NASA.)

The same alarm came on 15 seconds later, and this time Mission Control responded GO immediately. Three minutes later they were below 3,000 feet (900 m), descending about 1,000 feet (300 m) every 20 seconds when they had three alarms in about 30 seconds – a 1202 and two 1201s. Although Mission Control gave the GO signal without delay, we can well imagine Armstrong and Aldrin being pretty stressed by this so close to the ground. Armstrong later explained that in the simulations the mindset was that if you got a serious alarm you practiced aborting the mission – firing up the engine and returning to the CSM without landing. That was the purpose of simulations – practicing emergency maneuvers.

Armstrong explained when he got back to Earth that "In simulations we have a large number of failures and we are usually spring-loaded to the abort position. And in this case in the real flight, we are spring-loaded to the land position."

Later he explained further that "the concern here was not with the landing area we were going into but, rather, whether we could continue at all (because of the [computer] alarms). Consequently, our attention was directed toward clearing the alarms, keeping the machine flying, and assuring ourselves that control was adequate to continue without requiring an abort."

Aldrin later seemed to admit responsibility for causing the alarms, or at least the 1202 ones (see Fig. 4.4). In the 2006 documentary movie *In the Shadow of the Moon* he said that he switched on the rendezvous radar during the descent even though that was not called for in the flight plan. All the simulations had assumed that it was switched off, and so the computer overload it caused hadn't cropped up. Aldrin's thinking was that if they suddenly had to abort the landing and return to the CSM they would need the rendezvous radar to be working. He was somewhat paranoid about not being able to find the CSM again and recognized that the rendezvous radar would be essential to doing so.

They were now below 2,000 feet (600 m), and Armstrong took stock of the ground below and didn't like what he saw. "As we approached the 1,500-foot point, the program alarm seemed to be settling down, and we committed ourselves to continue. We could see the landing area…just short (and slightly north) of a large rocky crater surrounded with the large boulder field with very large rocks covering a high percentage of the surface." Only Armstrong saw the boulders because Aldrin was focused on the display panels and reading out the numbers on it to Armstrong, like height, speed and fuel remaining. The video of what Armstrong saw has been witnessed by millions of people since then, but at the time this was not available to Mission Control, so they were in the dark as to why Armstrong didn't land as planned. The rocks were up to 10 feet (3 m) in size and would have caused problems for the Lunar Module. So at about 500 feet (150 m) altitude Armstrong takes over control of the Lunar Module from the computer and manually drives it sideways at about 40 miles per hour (65 kph) and leveling out 300 feet (90 m) above the ground. They continue for about 30 seconds and then land on a spot that seems smooth. They had about another 40 seconds worth of fuel left.

The landing was smooth – "more like settling than landing," said Aldrin. They had chosen an almost perfectly flat landing site, with the Lunar Module tilting only by 4 degrees from the vertical – a smaller tilt than that of any of the Apollo missions that followed, with the 11 degree tilt of *Apollo 15* especially noticeable (see Fig. 4.5). Armstrong noted that the Lunar Module was a delight to fly – "just like a plane."

"Houston, Tranquility Base here; the Eagle has landed." Back in Houston it was 17 minutes and 40 seconds after 12 noon on Sunday July 20, 1969. Not for the last time, Neil Armstrong had come up with a phrase that would make the headlines.

Fig. 4.5. *Apollo 11* set down on this relatively flat area in the Sea of Tranquility, unlike *Apollo 15* two years later (inset), which ended up titled about 10 degrees off vertical in the Sea of Showers near a feature called Hadley Rille. (Illustration courtesy of NASA.)

Still tense from the alarms and hazards of the descent, the crew preferred to get on with the exploration of the Moon's surface rather than bed down and sleep, as called for in the flight plan – "bed down" as in Aldrin lies on the floor curving around the mound of the ascent engine while Armstrong sits on the cover of the ascent engine (see Fig. 4.6), his back against the wall and his legs supported by a strap he had jury-rigged to a bar. The Moon's gravity is about one-sixth that of Earth's, so the crew weighs that much less, and the thinking was that they could therefore endure the uncomfortable sleeping position. In the later *Apollo 16* mission the crew took the advice of the Houston controllers and tried to sleep before going outside – and regretted it. *Apollo 16* astronaut Charlie Duke said his "mind was just racing…..whirling. We had so much adrenaline pumping we could have gone two days without any problem."

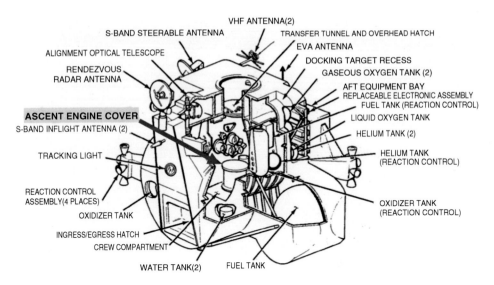

APOLLO LUNAR MODULE - ASCENT STAGE

Fig. 4.6. Diagram of the 7½-foot wide, 3½-foot deep (2.3 m by 1 m) crew compartment of the Lunar Module showing the drum-shaped cover of the ascent engine sticking up in the middle of the floor. Sleeping arrangement: Armstrong on top of the engine cover, Aldrin wrapped around the base. (Illustration courtesy of NASA.)

They had donned their spacesuits back in the CSM and would wear them until they got back there – the suits were bulky and awkward to don, and the tiny space inside the Lunar Module made it impractical to use it as a dressing room. The suits weighed a hefty 180 pounds or so (80 kg) on Earth, but in orbit of course they were weightless, and on the Moon they weighed just 30 pounds (16 kg) – the Moon's gravity being less than Earth's. In the Lunar Module the only dressing up needed was to put on the helmet and the gloves; then they were good to go walk on the Moon. One advantage of wearing the suits inside the Lunar Module was the protection it gave against micrometeorites – the wall of the Lunar Module was only the thickness of two or three layers of kitchen foil, barely 1 1/100 of an inch (¼ mm) thick. But the suits were bulky.

The bulk of the suits had made it a tight squeeze to get into the Lunar Module from the CSM, and once on the Moon it made it difficult to get out onto the ladder. Aldrin had to talk Armstrong through like a blind man – "Put your left foot to the right a little bit" and so on. Eventually he was able to announce that he was on "the porch" as they quaintly (mockingly?) described the top rung of the ladder. Then a few careful steps down and a final short hop off the lowest rung, and Armstrong "had joined the ranks of the forever quoted" [7] – "That's one small step for a man, one giant leap for mankind!" (see Fig. 4.7) It was Monday, July 21, 1969, 9:56 a.m. Houston time.

Fig. 4.7. A still from the TV footage watched as it was happening by hundreds of millions on Earth shows Armstrong step off the last rung of the ladder and on to the Moon's surface, where he remarked "That's one small step for a man, one giant leap for mankind." (Illustration courtesy of NASA.)

For those of us back on Earth perhaps the most amazing thing about this whole affair was that we could watch it live on TV as it happened, pretty much anywhere in the world. To his credit, although President Kennedy's commitment to go to the Moon back in 1961 ("before the decade is out") had been driven by relatively short-term political considerations, he had justified it at least in part by noting "the impact of this [space] adventure on the minds of men everywhere, who are attempting to make a determination of which road they should take." And in that same speech he had put his money where his mouth was concerning this propaganda objective by funding a program to "accelerate the use of space satellites for worldwide communications." Those communications satellites had duly come to fruition during the 1960s and were in place to broadcast Armstrong's "small step" to anywhere in the world that was interested.

It was the largest TV audience for any event at the time – 600 million viewers is the estimate. Politics dictated whether you saw it or not, so the Chinese weren't told that a man had walked on the Moon, let alone shown it on television. Nor were the citizens of North Korea, North Vietnam and Albania. In Moscow the news was broadcast without TV footage in a break in a volleyball match between two local teams.

Elsewhere the impact was huge! The message was "Eat your heart out Soviet Union; not only has the U. S. of A. gotten there first but we had the balls to show it on prime time television as it happened." It was a good news story at a time when bad news was all too common – the Vietnam War, riots in the ghettos, assassinations of political leaders. Kennedy's aim of impacting world opinion had been achieved with a vengeance. We will pass over the unfortunate irony that the late president's younger brother, Senator Ted Kennedy, stole some of the media headlines due to being involved in a car accident that Friday (July 18) in which his passenger Mary Jo Kopechne had died, then fled the scene and didn't report the accident until ten hours later.

The astronauts themselves (and their wives) got a taste of the impact during their five-and-a-half week whistlestop world tour that fall. More than 100 million people saw them in person during the stops in 23 countries: four in Latin America, eight in Western Europe, one in Eastern Europe (Yugoslavia), one in Africa, seven in Asia and two in Australasia. This was diplomacy with a capital "D" celebrating American ingenuity, initiative and science as never before. And wealth! Apollo cost a lot of money, and the United States could afford it while continuing to do the other things it did well, like build jet aircraft, live in air-conditioned luxury, and drive millions of motor vehicles on tens of thousands of miles of freeway. The American way of life was not only comfortable, it was far and away the most technologically advanced in the world.

"Returning Them Safely to Earth"

It was anything but trivial for Armstrong and Aldrin to get safely back to Earth, as promised by President Kennedy in 1961, where they could enjoy all the rewards and adulation that awaited them.

First of all they had work to do. Aldrin joined Armstrong on the surface, helped by Neil's guidance to negotiate the tricky exit from the Lunar Module ("Your toes are about to come over the sill"). The TV camera that took the grainy black and white images of the "small step" seen by the world had to be moved to give a view of the astronauts as they worked outside in their spacesuits. Moon rock had to be collected, and various objects had to be placed on the Moon's surface (see Fig. 4.8).

48 The Eagle's Journey

Fig. 4.8. (*Left*) "We came in peace for all mankind" says the plaque that Armstrong attached to a leg of the Lunar Module. (Illustration courtesy of NASA.) (*Right*): It's a tight squeeze for Aldrin getting out of the Lunar Module. (Illustration courtesy of NASA.)

Over 47 pounds (21 kg on Earth) of Moon rock was picked up and stored in special containers, but not before they had planted an 8-foot pole in the surface with a 5-foot by 3-foot stars-and-stripes flag attached. This provided a hugely symbolic backdrop to their congratulatory phone call from President Nixon – with the world listening in and watching.

Armstrong kept hold of the Hasselblad camera that provided the beautiful color photographs the world has come to associate with the mission. Sadly this means that there are no good photos of Armstrong himself on the surface, just ones he has taken of Aldrin.

But what photos they are! Especially the full length portrait of Aldrin, with Armstrong and the Lunar Module reflected in his helmet visor. His "power stance" has been popular in assertive portraits at least since the famous Hans Holbein portrait of Britain's King Henry VIII four hundred years earlier, which shows a vigorous Henry staring straight at the viewer as if to say "I rule here" (see Fig. 4.9). It was considered then (and still is now) to be a superb example of state propaganda. King Henry made many copies to distribute to those he wished to influence and encouraged his underlings to have copies made for themselves. In the Aldrin portrait, the astronaut's casual version of the pose suggests the completeness of the American triumph – not only have we come to the Moon, we had the foresight to pack a state of the art camera and found time to use it to record our achievements.

Fig. 4.9. Buzz Aldrin on the lunar surface. Reflected in his visor are Armstrong and the Lunar Module. Inset (*top*) is a copy of Holbein's 16th century painting of Britain's Henry VIII in a similar propaganda-style "power stance." Inset (*below*) is Aldrin's boot print on the lunar surface. (Illustrations courtesy of NASA and Walker Art Gallery, Liverpool.)

Another iconic image is the footprint, the imprint of Aldrin's boot (Fig. 4.9), which NASA still uses to represent its greatest historical achievements. Without the effects of wind or rain it will remain unchanged for millions of years. It brilliantly illustrates with a human touch the lifeless nature of the Moon and the fact that humans have been there to take its measure, in a way that no flag or piece of machinery can.

Two hours after the "small step" it was time to load up the Lunar Module with the rocks and get back inside. The crew also loaded up a strip of aluminum foil that had been left exposed to sunlight for an hour or so. The foil would bear the imprint of atomic particles coming from the Sun (the "solar wind") without being influenced by Earth's magnetic field. Three hours later, and about 21 hours since they had woken up that morning in the CSM, they tried to get some sleep.

Armstrong later explained that they kept their helmets and gloves on while resting to keep out dust and noise. Moon dust clings to the spacesuits because everything is so dry, and then once inside the Lunar Module the dust starts to get kicked around. The noise came from various pumps that circulated oxygen and heat, ran the carbon dioxide scrubbing machine, and so on. Armstrong, who had his head leaning against the cabin wall, found the noise especially hard to ignore. The light was also distracting, coming from the Sun, right through the apparently inadequate window blinds, and from the illuminated switches and displays on the instrument panel. Despite attempts to control it, the temperature fell as the night wore on, leveling off at about 61° F (16° C), making it impossible to sleep.

Next day (Houston time) the Lunar Module crew prepared to leave the Moon. They would leave behind the 2.7-ton[4] lower half of the Lunar Module (see Fig. 4.10), returning to the CSM traveling in the upper part – the so-called ascent stage. After twenty-one and a half hours on the surface they fired the ascent engine for just over seven minutes, putting them back into orbit. On the last three of the later Apollo missions (15, 16 and 17) we could see the takeoff from one of the cameras they had left behind – the spacecraft rises rapidly and steadily, not at all like the slow, ponderous rise of the Saturn V at Cape Kennedy. We also see the dust kicked up by the blast from the engine, momentarily obscuring the left-behind lander, flag and other artifacts.

Michael Collins, the astronaut left alone in the CSM while Armstrong and Aldrin flew off in the Lunar Module, was upbeat about the experience. He did note somewhat wryly that he was probably the only American who hadn't watched live TV of his colleagues walking on the Moon's surface. The CSM's technology wasn't up to receiving TV pictures from either Earth or direct from the Lunar Module. But generally he had plenty to keep himself occupied, updating the computer, clearing the cabin so that the Moon rocks could be stored away and other housekeeping tasks. He also spent time unsuccessfully looking through the CSM's small telescope trying to spot the Lunar Module on the surface – Eagle was about 4 miles from its intended landing site, and Collins never found it. He was rarely in contact with Houston Mission Control partly because about a third of the time he

---

[4] There are at least three different weights that are pronounced "ton". For brevity, I use the word "ton" to signify a weight of 1,000 kg (about 2,205 lbs) instead of "tonne" or "metric ton". Note that in the United States and Canada, "ton" usually means 2,000 lbs, while in the rest of the world it usually means 2,240 lbs.

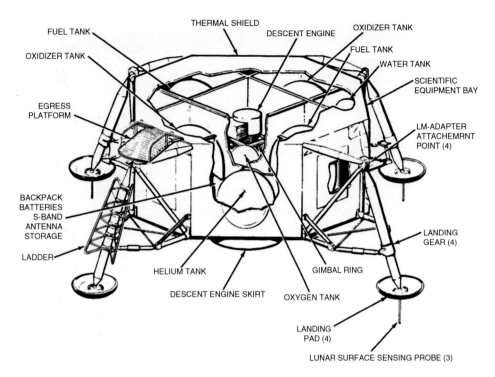

Fig. 4.10. Left behind on the Moon's surface: the descent stage of the Lunar Module including the spindly legs, together with the American flag and various scientific instruments. (Illustration courtesy of NASA.)

was on the far side of the Moon and thus out of sight of Earth, and partly because there was only one link between Earth and Moon, and the men on the surface used that most of the time.

One of many fears before the flight was that the ascent engine wouldn't work, stranding the astronauts on the Moon with no hope of rescue. This potential problem didn't seem to concern the crew, no doubt partly due to their backgrounds as test pilots. Aldrin wisecracked that "We're number one on the runway," and once aloft that "We're going right down US-1," while according to the NASA doctors Armstrong's heart rate was absolutely normal during the ascent.

In any case the vitriolic, toxic chemicals in the two fuel tanks mixed together and explosively drove the spacecraft off the surface, burning about 2 tons of the fuel for seven minutes until they reached almost 4,000 miles per hour (6,400 kmph) and were in orbit heading towards the CSM. "A beautiful fleeting final view of Tranquility Base as we lifted up and away from it," reported Armstrong later, adding that the it was "very smooth; a very quiet ride". The rendezvous radar was switched off for the early part of the journey so as not to overload the Lunar Module computer and generate the scary alarms they had experienced during the landing.

52   The Eagle's Journey

Fig. 4.11. The 2¼-ton ascent stage of the Lunar Module returns to Michael Collins watching through the window of the CSM. Originally the Lunar Module weighed over 15 tons but had burned up 8½ tons of fuel while landing, nearly 2½ tons on the return, and left the 2-ton descent stage on the surface. The link up with the CSM is witnessed by Mission Control back on Earth, which is just coming into view. (Illustration courtesy of NASA.)

For the audience back on Earth, tension was increased by the saga of the Soviet *Luna 15* probe launched three days before *Apollo 11* and in orbit around the Moon throughout Armstrong's and Aldrin's stay on the surface. The official Soviet story was that *Luna 15* would carry out surface photography and other scientific tasks, but we now know that it was intended to return material from the Moon's surface to Earth, just ahead of the return of *Apollo 11*. It was a last gasp effort by the Soviets to compete with the United States. The hope was to bring back lunar materials a few days before *Apollo 11* or even better, to demonstrate the benefits of unmanned lunar exploration if *Apollo 11* failed in some way.

There was much speculation in the western media as *Apollo 11* progressed, wondering if *Luna 15* might somehow get in the way of the returning Lunar Module. Mission Control and the *Apollo 11* crew were in fact kept relatively well informed by the Soviets about the orbit of *Luna 15* helped by the presence in Moscow of *Apollo 8* astronaut Frank Borman on a semi-official visit who had liaised with Soviet space officials.

Luna 15 was successfully placed into orbit around the Moon, but the Soviet controllers had great difficulty in determining and predicting its orbit due to the Moon's irregular gravity field. After several postponements they fired the descent engines just two hours before Armstrong and Aldrin were due to take off but smashed into the side of a lunar mountain about 500 miles (800 km) east of Tranquility Base. Ironically because of the delays, even if *Luna 15* had landed, scooped up some soil and returned it to Earth, it would have got back two hours after *Apollo 11* splashed down.

As an aside, July 1969 was an especially bad month for Soviet lunar ambitions. We will discuss this in more detail in Chapter 8. On July 3, the second test launch of their N1 giant rocket (their equivalent of the Saturn V) blew up 15 seconds after liftoff at a height of 700 feet (200 m) in a gigantic explosion that destroyed the launch pad, broke windows 25 miles (40 km) away and scattered debris for more than 5 miles (8 km), including a fuel tank weighing 0.4 tons that landed on a building 4 miles (6 km) from the pad. Fortunately there were no human casualties, since the many assembled dignitaries were safely in bunkers 4 miles away [8]. This event was kept secret from the west (and from the Soviet public) for the next 20 years.[5]

Unaffected by events in the Soviet Union, the Lunar Module went twice around the Moon as it gradually adjusted its orbit to pull up slowly to the CSM. Thus it was nearly four hours after leaving Tranquility Base that they came in view of their ride home – the CSM and its chauffeur, Michael Collins (see Fig. 4.11). Collins said that he didn't feel the two craft docking, so gentle was the approach. But he certainly felt the effect of firing the bolts that were intended to lock the two craft together – "All hell broke loose" was his remark, as the slightly off balance pair of spacecraft pulled tight together rocking and twisting back and forth for eight or nine seconds until they were safely bound together.

More housekeeping chores were now required, including Armstrong and Aldrin vacuuming the Moon dust off each other's suits that otherwise would float around in the zero-gravity and get into eyes, ears, mouths, food, drink, equipment – everything. Then they could pass the precious Moon rocks through the hatch to Collins before entering themselves and casting the Lunar Module adrift.[6] It was a further three and a half hours before they fired up the engine of the CSM to take them out

---

[5] Rumors of the explosion including the incorrect report of large numbers of human casualties were published in the West a few months later.

[6] It is thought to have been gradually pulled in towards the surface by the Moon's irregular gravity and to have crashed within a few months. The crash site has never been found. Later Apollo missions fired the engines to drive the empty ascent stage into the surface at a pre-selected time and place so that its impact could be measured by the seismic instruments left behind on the surface by the astronauts.

of lunar orbit and back towards Earth. The engine burn took place while they were on the far side of the Moon and out of touch with Mission Control. There were nervous faces in Houston until the now fast-moving CSM appeared around the edge of the Moon, heading for home, rotating steadily all the way to avoid overheating (the BBQ roll mentioned earlier).

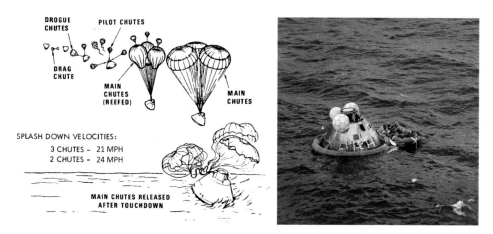

Fig. 4.12. (*Left*) Safely through the violent and fiery re-entry to Earth's atmosphere, the heat shield is jettisoned at 24,000 feet (7,300 m) altitude and the parachutes deployed as shown. (Illustration courtesy of NASA.) (*Right*) A U. S. Navy sailor helped the crew into the life raft he had brought by helicopter from the aircraft carrier *Hornet* 14 miles away. (Illustration courtesy of NASA.)

One last dangerous moment remained – the re-entry into Earth's atmosphere two and a half days later. The material used as a heat shield for long-range missiles was okay for returning from orbiting Earth at 18,000 miles per hour (5,500 kmph), and this was the technology for the Mercury and Gemini missions. But the 24,700 miles per hour (7,500 kmph) return from the Moon gave rise to too much heat – about 5,000° F (2,760° C), so a new approach was taken for the Apollo missions – a coating[7] that gradually burned off ("ablated" was the technical term), lasting long enough for the craft to slow down sufficiently to deploy its parachutes (see Fig. 4.12). *Apollo 8* and *Apollo 10* had already proven the technology, but still the dynamics of slowing from 24,000 miles per hour at 400,000 feet (120 km) altitude to less than 1,000 miles per hour in under 2 minutes are very severe, and the crew had to cope with forces of more than about 6 g at two points in the descent – once as they bottomed out at 180,000 feet (55 km) altitude and bounced back up to 200,000 feet (60 km), then again at 120,000 feet (37 km). Thirty-four years later

---

[7] The coating was made of phenolyic epoxy resin and was applied to a brazed stainless steel honeycomb structure.

we would see the result of a failed re-entry when the space shuttle *Columbia* disintegrated with the tragic loss of six astronauts because its re-entry shield was damaged.

Apollo 11 had no hiccups with the main re-entry, although the capsule did end up upside down in the ocean due to rough seas and strong winds – undignified perhaps but perfectly safe since it was designed to float either way up.[8] The Service Module (the "S" in CSM) having been jettisoned 15 minutes before re-entry, the weight of the returned Command Module was just 5½ tons, all that was left of the 3,000 tons they had climbed aboard 8 days earlier. It was 22:50 Houston time on July 24, 1969.

President Nixon was there in the middle of the Pacific Ocean to greet them on the aircraft carrier U. S. S. *Hornet*. Mission Control had moved their splashdown point 250 miles (400 km) east and closer to Hawaii due to a thunderstorm in the planned landing area. This made the helicopter ride for President Nixon from Hawaii a bit shorter, but the astronauts didn't get to Hawaii any faster because the captain of the U. S. S. *Hornet* steamed at a slower rate, taking the full planned 55 hours to reach port. The Moon rocks and the rolls of film taken by the *Apollo 11* crew got the VIP treatment and were flown by helicopter and then fast jet all the way from the ship to Houston – split between several helicopters and jets so that they wouldn't all be lost if one crashed.

President Nixon somewhat overstated the significance of the mission in his public remarks on the *Hornet,* calling it "the greatest week in the history of the world since the creation," blithely ignoring some other events in the past of significance to Christians, Muslims and others.

The TV cameras were on the recovery ship, too, beaming the pictures to millions of admiring fans around the world. But the ticker tape parades, even the greetings to wives and children, had to wait for two weeks while the crew went into quarantine in case they had contracted some hazardous disease during their travels; they were accompanied by a doctor and other medical staff, a chef, two stewards, a photographer and a press officer – a total of 12 men to look after the

---

[8] This happened on about half of the Apollo re-entries. On Apollo 8 Ed Anders recalled "hanging upside down in the ocean with all the garbage falling on us" until the three flotation balloons righted the capsule while his Commander Frank Borman was "sick as a dog from seasickness" during the 43 minute wait in the 10' (3m) swell until frogmen arrived. The most serious incident was the re-entry of the 1975 Apollo-Soyuz joint mission with the USSR, caused on this occasion by failure of one of the three parachutes to open. Toxic gas from the fuel tanks seeped into the cabin during the descent, incapacitating the crew, and the upside down position led to some delay in getting them out. Vance Brand had to be resuscitated by his two astronaut colleagues, Tom Stafford and Deke Slayton, before the hatch was opened. All three were put in intensive care but soon recovered.

three astronauts. Then[9] off they set to tumultuous crowds across the United States, in New York, Chicago and Los Angeles, and then 23 countries in a month and a half. Their new careers as unofficial American diplomats had begun, whether they liked it or not.

In a moment we will take a quick look at the later Apollo missions, then briefly assess their legacy – scientific, technical, political, social and economic. Then we will consider the Soviet Union's attempt to win the Moon race, before turning to consider why Cernan and Schmidt on *Apollo 17* were the last humans to step onto the surface of the Moon to this day, almost 50 years later. But before all of that let's take a quick look at some of the other achievements of Apollo, since the scale of the engineering has been a stumbling block for later would-be Moon visitors.

References

1. Orloff, R. W., *Apollo by the Numbers*, NASA SP-2000-4029, 2000 (history.nasa.gov/SP-4029.pdf).
2. Anon, *Apollo 11 Press Kit*, July 6, 1969 (www.hq.nasa.gov/alsj/a11/A11_PressKit.pdf).
3. Ezell, L.N., *NASA Historical Data Book, 1958-1968, Vol 2: Programs & Projects*, NASA SP-4012v2, 1988 (history.nasa.gov/SP-4012/cover.html).
4. Van Nimmen, J., Bruno, L. C., Rosholt, R. L., *NASA Historical Data Book 1958-1968, Vol 1: NASA Resources*, NASA SP-4012v1, 1976 (history.nasa.gov/SP-4012/cover.html).
5. Various, *The Apollo Spacecraft – A Chronology*, NASA SP-4009, 1969/1973/1978 (history.nasa.gov/SP-4009/cover.htm).
6. Konopliv, A. S., et al, A High Resolution Lunar Gravity Field and Predicted Orbit Behaviour; Paper AAS 93-622, August 1993.
7. Mailer, N., *A Fire on the Moon*, Pan Books (London), 1970, p. 364.
8. Siddiqi, A.; *The Soviet Space Race with Apollo*, University Press of Florida (2003) pp. 688-697.

---

[9] One local assignment for Armstrong before he left Houston was to hand out Apollo Individual Achievement Awards to several of my TRW colleagues and me at our office just outside Mission Control – in my case for the work my team had performed to provide the correct longitudes of a third of NASA's ground stations that had been shown during *Apollo 8* to be seriously in error.

5

American Knowhow

In the previous chapters the Saturn V rocket has been presented as the greatest technical triumph of the program. We have also been able to appreciate the exquisite design of the spacecraft carried aloft by the Saturn V: first the Command and Service Module (CSM) that led to the death of three astronauts in 1967 before it had ever been into space, and which when safely redesigned took the astronauts to and from the Moon and into orbit around the Moon; then the fragile-looking Lunar Module spacecraft, which actually touched down on the Moon's surface providing a home for two astronauts for a day or so and then brought them back to the CSM waiting in orbit above.

Three other engineering challenges faced by the Moon landing team were also groundbreaking at the time and are worth mentioning when we come to consider what it takes to land humans on the Moon:

- the launch complex at Cape Canaveral, including the largest building in the world.
- the network of tracking and communication stations and computer facilities needed to monitor and control the mission.
- the management of the whole program, with its technical, political, industrial, human and financial challenges.

This list excludes a host of other items, such as the spacesuits worn by the astronauts while walking on the Moon. These spacesuits were essentially autonomous spacecraft in their own right. Also excluded is the two-man Gemini spacecraft whose twelve missions in 1965 and 1966 were trailblazers for Apollo, enabling NASA to fine tune technologies and procedures necessary for going to the Moon. The five unmanned lunar orbiter spacecraft were also critical to Apollo's

58 American Knowhow

success, mapping the Moon in unprecedented detail, front and back, and feeling out the full measure of the Moon's gravity field. But the Saturn V, the CSM, the Lunar Module and the three listed above are singled out as the main challenges that the United States overcame sooner than the Soviet Union, thus winning the race to land men on the Moon.

Cape Canaveral – Stage Zero

This was the largest enclosed space in the world on top of sandy soil in a place periodically swept by hurricanes.

Von Braun's approach with the V2 rockets during World War II was to build them in a hangar – a low building with people working around it, then take them to the pad when ready to launch. But the Saturn V was so big (360 feet/110 m tall) and fragile that it was impossible to move it from horizontal to vertical – its structure would buckle because its thin walls were designed to support its weight when upright.[1]

So they needed a building higher than the 360 feet in order to have roof-mounted cranes that could erect the various stages. And there could be four in various stages of assembly at the same time. Thus arose the enormous Vehicle (or Vertical) Assembly Building (clouds don't form inside it – urban legend), an immense rectangular building with walls 525 feet (160 m) high and thousands of piles going 160 feet (50 m) the other way down to the bedrock that prevent it from being blown away by the Florida wind.

The floor area covers 8 acres (more than 3 hectares) – enough for six football fields. Each of the doors is 456 feet (139 m) high – and wide enough to move the U. N. headquarter building through it.

The Saturn V then had to be moved from the Vehicle Assembly Building to the launch site 3½ miles (5½ km) away, standing upright (see Fig. 5.1). The technology was not too different from that required for a giant stripmining shovel – a self-leveling machine that carried enormous weights moving slowly on tank treads. The Saturn V needed a bigger and better version, but the idea was well proven. Alternatives such as digging a canal to the launch pad so as to move the rocket on a barge were quietly discarded.

The 380-foot (115-m)-high umbilical tower was another highly visible part of the launch complex. But then everything for the Saturn V was gigantic – and complicated. The cryogenic fuel had to be stored and then loaded into the tanks in the hours before launch. The flames of the first stage had to be deflected by a huge

---

[1] Unless otherwise stated, factual information about the Apollo program comes from [1–5].

Fig. 5.1. May 20, 1969, the 363-foot (110-m) -high Saturn V with *Apollo 11* inside the cone near the top (without crew) crawls at 1 mile per hour for 3½ miles from the Vertical Assembly Building (left) to the launch pad on the coast. (Illustration courtesy of NASA.)

iron construction in a trench beneath the rocket. Otherwise they would bounce back up and destroy the base of the rocket.

It was not unreasonable for the engineers who constructed this set of monstrous machines and structures to refer to it as Stage-Zero of the rocket.

Stations and Computers

Radios, radars and computers were and are essential parts of any space mission. There is no alternative to radio for communicating with a spacecraft, whether to speak to the crew, to get information on the state of the vehicle or to send information and instructions to the vehicle. The missions that came before Apollo had enabled NASA to roll out radio and radar facilities around the world, and to evolve the equipment to take account of advances in technology and to learn from earlier missions.

The robotic probes to the Moon, Mars and Venus had been particularly helpful in establishing how many stations were needed for a mission that went beyond the Earth into deep space. The bare minimum was three, spaced around the globe – California, Australia and Spain were the locations used by NASA. Note that one of those (Australia) is in the southern hemisphere. It helps not only to have stations separated in longitude (east to west) but also separated in latitude (north to south).

NASA added a network of other stations that gave near continuous coverage of the launch phase of a mission, that is to say stretching out into the Atlantic southeast of Florida, and also filling in the wide gaps between the three core stations. These were on isolated islands like Hawaii and Guam in the Pacific, and Ascension Island and Grand Canary in the Atlantic,[2] as well as at continental sites on four continents.

Radio and radar had been widely used during the Second World War and after that in the Korean War. Computers, though, were comparatively new. The use of all of these technologies for space missions benefited from the systems developed by the U. S. military from the mid-1950s onwards for long-range missiles. The military also drove forward research into Earth's gravity field, whose irregularities had to be known in detail in order to guide a long-range missile to its target – information that was also required for space vehicles.

The computers of Mission Control in Houston were the best that money could buy at the time, but then as now a computer was only as good as the software that had been programmed into it. One company was synonymous with the word "computing" in the 1960s – IBM. Many other companies made computers and the associated software, but IBM dominated the market throughout the world. The relative success of the other main computer suppliers was illustrated by the commonly used phrase "IBM and the seven dwarfs" to describe their relationship to the market leader.[3] It will come as no surprise to hear that IBM supplied not only most of the computers and software in Mission Control but also the computers and most of the software inside the Saturn V, the CSM and the Lunar Module. Many other companies provided additional expertise so that the software used by Mission Control worked reliably and efficiently.

---

[2] The longitudes of these island sites were shown to be erroneous during the *Apollo 8* mission in December 1968. I led a small team at TRW in Houston to provide accurate values in time for *Apollo 10* in May 1969.

[3] The seven were Burroughs, Control Data Corporation, General Electric, Honeywell, NCR and UNIVAC.

Fig. 5.2. Apollo Mission Control in Houston. The men (and a very few women) in Mission Control were the human face of the worldwide network of radar devices, communications stations, computers and software that sent instructions to the astronauts. In the picture staff relax at the end of the *Apollo 11* mission on July 26, 1969. On the rightmost wall screen, President Richard M. Nixon, on the USS *Hornet* in the Pacific Ocean, is interviewing Armstrong, Aldrin and Collins, who are behind the glass of the quarantine facility. The blue central wall screen proclaims "Task Accomplished!"

The media tended to focus on the staff at Mission Control (see Fig. 5.2), but the reality was that they relied almost totally on the software in those IBM computers to calculate trajectories and the like. So the breakthrough was due not the caliber of the mission controllers (excellent though that was) but to the sophistication of the software and the computers.

Managing the Program

It's easy to spot when management is bad. And there's plenty of bad management about! Delays and cost overruns in big projects abound or computers are hacked again and again. This was true in the 1960s, and then as now it was often difficult to decide where the blame lay.

It's less easy to recognize and explain good management. Successful completion of the Apollo program within the time limit imposed by President Kennedy called for good management at all levels. There were of course examples of bad management along the way, not least the failings at North American Aviation and in NASA that led to the deaths of Chaffee, Grissom and White in 1967. But there were also examples of great management.

Three factors illustrate the high quality of the management of Apollo and show how it involved both political and technical issues. First was the decision to rendezvous in orbit around the Moon instead of in Earth orbit that has already been discussed (Chapter 3). Second was the management of the politics and third was the aggressive test program. Let's explore these last two.

NASA was created in 1958 to manage all U. S. government-funded civilian space programs, and this action by President Eisenhower should be recognized as critical in ensuring that the Apollo program was managed so well. Contrast this with the arrangements in the Soviet Union where there was no equivalent to NASA, and programs were approved and managed by a mixture of personal authority and random political decisions. This Soviet management approach worked sufficiently well to get *Sputnik* and Gagarin into orbit, but then the wheels came off, as will be explained in Chapter 8.

The choice of James Webb as NASA's second top administrator was criticized at the time as being too political. It *was* a political appointment, as is often the case for the heads and other senior managers of all U. S. government departments and agencies. So when President Kennedy was elected, the existing administrator, T. Keith Glennan, handed in his resignation, as did every other Eisenhower-era appointee, and the new president got to choose his replacement. But it proved difficult to find anyone willing to accept the job. According to Vice President Johnson the first seventeen candidates offered the job turned it down. One official privy to the process reckoned that the high refusal rate was because Johnson would personally contact a candidate, and the thought that they would effectively be working for the sometimes abrasive Johnson turned a lot of people off.

The 18th choice was James Webb, who was an old-style politically astute industrialist and lobbyist from a Southern state (North Carolina). This made him very different from the young, intellectual high flyers being appointed by Kennedy to other positions, but closer in style to that of the vice president.

Webb proved to be highly effective at persuading Congress to fund the programs that Kennedy wanted. The scale of the funding needed for the Apollo program was daunting. From 1964 to 1967 NASA's budget (dominated by Apollo) represented about 4% of the federal budget. By 1975, after the Apollo program had ended, NASA's budget had dropped back to its 1961 level of 1%

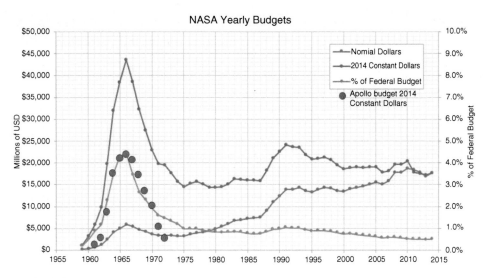

Fig. 5.3. NASA's budget (orange line) was dominated by the Apollo program (orange dots) from 1963 to 1970. The orange values are adjusted for inflation to 2014. The blue line is current year dollars. The gray line shows NASA's proportion of the federal budget (right-hand scale) peaking at about 4% in the mid-1960s then gradually falling to today's 0.5%

and trickled on down thereafter to about half of one percent, where it remains today (see Fig. 5.3) [6].

Webb showed great skill in finessing the Washington, D. C., system so that the politicians, the public and the industrial lobbyists remained onside as Apollo went from bright idea to expensive realization. Whereas one might have expected the Department of Defense to object to funds going to NASA that might otherwise have come its way, Webb persuaded the analytically minded Defense Secretary Robert McNamara to lend support to the Moon-landing program on the grounds that it would give the aerospace industry badly needed work without coming out of the defense budget [7].

The second key figure in the Apollo management story is George Mueller. He was in charge of all human spaceflight programs at NASA, and deployed a cold-eyed analytical approach to the task of making the Gemini and Apollo programs a success. Mueller[4] was born in St. Louis, MO, and worked for Bell Labs during World War II on airborne radars, then for what became TRW[5] on electronic and missile programs. He identified early on that the schedule of placing men on the

---

[4] Pronounced "miller."
[5] Disclosure of interest: I worked for TRW on the Apollo Program 1967-1970.

Moon by 1970 was too tight to adopt the approach taken by von Braun or by the NASA staff, who had come from its aviation precursor, the National Advisory Committee for Aeronautics, NACA.

Both of these groups had a "build a little, test a little" philosophy, which was safe but slow. Von Braun and the rocket engineers in Huntsville had developed the first long-range rockets by learning from failure (after failure, after failure….). Their instinct was to build the first stage of the Saturn V and test it. In parallel do the same for the second and third stages. When each was working then put the first two together and test that. When that had been tested, proceed to add the third stage and test that.

The aviation people from NACA were driven by the need for aircraft to meet very high safety standards. Nothing less than 100 percent safe was their goal. What they and von Braun's team had in common was a love of engineering excellence. With excellence and safety as the all-consuming objectives both these groups ended up with systems that contained wide margins for error.

George Mueller came from the missile business – designing and building hundreds of missiles to carry nuclear bombs over long distances. The missile boys came from a background where you were asked to work out the best use of a billion dollars, say, to create a nuclear deterrent. One way would be to build a small number of exquisite missiles that would overcome any defense and thus be sure to reach their target but might be vulnerable to being knocked out in a surprise (Pearl Harbor-style) attack. Another way would be to build a large number of good missiles, a majority of which would hit their targets, but by their very number would be impossible to wipe out in a surprise attack. The relative merits of these and other scenarios could be analyzed and tradeoffs carried out. And this is the kind of system engineering analysis that Mueller and his team did for Apollo.

Some might say that the pragmatic approach of the missile boys was a cavalier or "cowboy" attitude, but there is solid philosophical backing for it. Two hundred years earlier the French philosopher Voltaire coined the phrase "The best is the enemy of the good [8]," by which he meant that seeking perfection is a fool's game and causes you to miss out on many good experiences. The philosophy can be seen in practice in the discipline of emergency medical triage, where treatment is withheld from a patient if limited medical resources can be used more effectively on someone else. For example, a patient with no hope of survival might be passed over in order to treat a less injured person who has a chance of survival if treated promptly.

System engineers argued that looking for the "best compromise" or "best balance" was the right way to spend taxpayers' money, not building some perfect machine that would be too expensive and arrive too late.

Helping Mueller to drive this philosophy into all parts of the program was another engineer with missile experience, Joe Shea, who took the notion of not testing Saturn V stages separately to its logical conclusion. In his scheme, first of all, you tested all stages of the rocket together so that if the first stage worked you then got to test the second stage, and if that worked then the third stage was also tested. Testing each stage separately assumes they will fail. Testing them together assumes they might work.

Even if von Braun and others accepted this (and they didn't), the next step in the logic really got their attention. If the Saturn V works a couple of times, you fly men on it on the next flight. The plan had been to have six or eight launches without any crew in order to give confidence in the rocket, but Mueller and Shea pointed out that to have any statistical significance you would need six or eight *dozen* flights. There was simply no time and not enough money to test the Saturn V in the way you would a new airplane – hundreds of hours of test flights in all sorts of different conditions before allowing it to be used. The best you could do was to trust the results of your ground testing (firing engines in static test stands and the like). If it worked when flown, you used it!

The importance of testing everything on the ground before you had a test launch was the key (see Fig. 5.4). You tested all the pieces together once you reached the point of launching them – "all-up-testing" was how they described it. We will see in a later chapter that the Soviets tried to save time and money by *not* testing the complete first stage of their giant rocket on the ground, and the result was failure.

At first Mueller met strong resistance to the all-up-testing approach. But the hard fact was that it was the only way to get men on the Moon before 1970. The relentless logic of having to meet the schedule gradually wore down the skeptics, so eventually the success-oriented testing strategy was used for Apollo. The first two Saturn V flights flew unmanned in November 1967 and April 1968, and although the second one revealed some problems, the third one was Apollo 8 that in December 1968 flew around the Moon with three astronauts onboard!

The same relentless logic was used to get acceptance of the idea of rendezvous in lunar orbit that was discussed in Chapter 3.

Shea left the program after the 1967 fire of *Apollo 1* that killed Chaffee, Grissom and White. The aggressive schedule was driven forward to the end by another veteran of the missile business that Mueller borrowed from the U. S. Air Force in 1964 to manage the whole Apollo program – General Sam Phillips, who previously had overseen the Minuteman ICBM program.[6]

---

[6] He returned to the Air Force a week after *Apollo 11* came home.

66   American Knowhow

Fig. 5.4. Ground testing the Saturn V engines. (*Left, top*) Testing a single F-1 engine. (*Left, below*) Testing five F-1 engines as in the first stage. (*Right, top*) Testing a single J2 engine as in the third stage. (*Right, below*) Testing five J2 engines as in the second stage. (Illustrations courtesy of NASA.)

Some would say that today NASA's human spaceflight program is allergic to risk and tending to add cost without achieving much. If so, it is out of step with the management style of the Apollo era.

References

1. Orloff, R. W., *Apollo by the Numbers*, NASA SP-2000-4029, 2000 (history.nasa.gov/SP-4029.pdf).
2. Anon, *Apollo 11 Press Kit*, July 6, 1969 (www.hq.nasa.gov/alsj/a11/A11_PressKit.pdf).
3. Ezell L N, *NASA Historical Data Book, 1958-1968, Vol 2: Programs & Projects*, NASA SP-4012v2, 1988 (history.nasa.gov/SP-4012/cover.html).

4. Van Nimmen, J., Bruno, L. C., Rosholt, R. L., *NASA Historical Data Book 1958-1968, Vol 1: NASA Resources*, NASA SP-4012v1, 1976 (history.nasa.gov/SP-4012/cover.html).
5. Various, *The Apollo Spacecraft – A Chronology*, NASA SP-4009, 1969/1973/1978 (history.nasa.gov/SP-4009/cover.htm).
6. https://en.wikipedia.org/wiki/File:NASA_budget_linegraph_BH.PNG accessed April 9th 2018, modified by the author.
7. Norris P, *Spies in the Sky*, Springer Praxis (Chichester, UK), 2007, p.16.
8. *The Oxford Dictionary of Quotations*, (3rd edition), Book Club Assoc. (London) 1981, p. 561.

6

After *Apollo 11*

NASA had planned for ten Moon landings – *Apollo 11* through *20*. However after the success of *Apollo 11*, the purpose of the remaining missions was increasingly questioned.

The second mission, *Apollo 12,* went ahead as planned on November 14, 1969, carrying an all-navy crew of Charles (Pete) Conrad, Richard Gordon and Alan Bean, with the primary goal of achieving a precision landing on the Moon – *Apollo 11* having ended up uncomfortably far from its intended landing site. The mission started with a bang as the rocket passed through clouds and was struck twice by lightning. The launch might have been postponed until the weather improved were it not for the presence of President Nixon at Cape Kennedy to witness it in person – the only Apollo launch watched by a president. The lightning knocked out the computer systems in the rocket, but the back-up computers kicked in and the mission proceeded without difficulty.

The plan was to land close to an April 1967 robotic probe, *Surveyor 3,* and thus demonstrate beyond question the ability to choose a spot and go there. *Apollo 11* had landed 15 miles (24 km) away from another such probe, the September 1967 *Surveyor 5,* but without any intention of Armstrong and Aldrin visiting it. The navigational difficulties that *Apollo 11* had experienced were avoided by *Apollo 12* due to Mission Control giving them last-minute updates to their trajectory based on how late or early they were appearing from behind the Moon.

Everything went to plan, and Conrad and Bean touched down within sight of *Surveyor 3* (see Fig. 6.1). Stepping down from the ladder to the Moon's surface the 5 feet 6½ inch (169 cm) Conrad said "that may have been a small step for Neil but that's a long one for me." Neil Armstrong was 5 feet 11 inches (180 cm). To their haul of Moon rocks they added pieces of *Surveyor 3,* including a TV camera and a soil scoop so that engineers back on Earth could study what two years of exposure on the Moon's surface had done to them. Their stay on the lunar surface lasted 31 hours and included two separate Moon walks. Their time on the surface was to have been broadcast in color, unlike the black and white of *Apollo 11,* but Bean accidentally pointed the TV camera into the Sun, destroying its picture-taking ability.

Fig. 6.1. November 1969. Pete Conrad inspects the *Surveyor 3* robotic probe that landed on the Moon in April 1967. The Apollo Lunar Module is 300 yards (275 m) away behind Conrad. (Illustration courtesy of NASA.)

The anticlimax following the enormous success of *Apollo 11* became evident when in January 1970 the last of the planned Apollo missions, *Apollo 20,* was officially canceled. The news had been unofficially released back in July as *Apollo 11* was on its three-day homeward journey from the Moon.

Public interest in Apollo was rekindled with the drama of *Apollo 13* in April 1970. Two days into the mission with the Apollo craft 200,000 miles (320,000 km) from Earth and accelerating towards the Moon an oxygen tank exploded in the Service Module section of the CSM. John (Rusty) Swigert first said "Houston,

I believe we've had a problem here," which an uncomprehending Mission Control asked to be repeated to which Jim Lovell replied "Houston, we've had a problem." The oxygen in the CSM rapidly leaked out into space, forcing the crew to transfer into the Lunar Module in order to survive the shortened but still three-and-a-half-day long journey around the Moon and straight back to Earth. They had to return briefly to the Command Module for the re-entry and landing.

The *Apollo 13* astronauts were hailed as heroes at the time, and this was unexpectedly repeated in 1995 with the success of the Hollywood version of the story starring Tom Hanks and Kevin Bacon.

The inevitable delays caused by analyzing the causes of the explosion and taking actions to avoid its repetition meant that there was a nine-month gap until the next Apollo mission. In the meantime, two more out of the originally planned ten missions were canceled, meaning that *Apollo 17* would be the last.

In January 1971 *Apollo 14* took astronauts Alan Shepard, Stuart Roosa and Edgar Mitchell to a hilly region of the Moon called Fra Mauro about 110 miles (175 km) southeast of where *Apollo 12* had landed. There was a surprisingly large crowd at Cape Kennedy to witness the launch, perhaps fearing that this would be the last Apollo mission and no doubt drawn by the warm Florida winter weather. The descent to the surface in the Lunar Module was as fraught as that of *Apollo 11,* with the crew having to deal with computer alarms and a faulty radar on/off switch during landing.

The landing was especially pleasing for the commander, Alan Shepard. He had been America's first man in space as a Mercury astronaut in 1960 (see Chapter 1), but had lost out in the fame sweepstake to his colleague John Glenn, who had been the first American to go into orbit a year later. Shepard's flight was suborbital, just a 15-minute up and down journey. The rivalry between Shepard and Glenn had been intense before either had flown, not helped by their different lifestyles. Glenn was the clean-cut religious family man at all times, Shepard was likewise when at home, but away from home was a different matter. He was a fast-living, hard drinking, fighter jock that the young ladies who frequented the beaches at the Cape found attractive. Glenn had lectured his six Mercury astronaut colleagues one time criticizing their fast living, and pointing out that they were now in the media spotlight, where discretion was required. Most of the other Mercury astronauts were annoyed by Glenn's moral pontifications, but Shepard most of all [1]. Shepard's satisfaction at being the fifth man, and the only one of the seven Mercury astronauts to walk on the Moon, was heightened further by the fact that he was refuting the critics who had said that at 47 he was too old.

The two Moon walks by Shepard and Mitchell were hard work. They had to pull a trolley loaded with instruments, cameras and film, containers for Moon rocks and the like that fully loaded could weigh 300 pounds (140 kg, Earth weight). They found it hard to pull across the dusty lunar surface, having to carry it up a slope on one occasion. This handicap was all the greater as they had to walk three-quarters of a mile (1.2 km) to a nearby crater during the second Moon walk, and in fact they stopped short 50 feet (15 m) from the crater rim, having overrun the time allotted.

Shepard wasn't content to complete the two Moon walks and decided to demonstrate to TV viewers on Earth how the Moon's low gravity could improve your golf game. With a makeshift golf club he hit two golf balls that he had brought onboard as personal items, one of which went 400 yards (365 m) into the distance. Mission controllers in Houston found this hilarious to judge by their roars of laughter, but ever after critics of the space program referred to the high cost of "sending a man to the Moon to play golf."

Fig. 6.2. The last time men stood on the Moon and looked back at Earth was during *Apollo 17*. This image was taken on December 12, 1972. (Illustration courtesy of NASA.)

The final three missions, *Apollos 15, 16* and *17,* were more sophisticated in that they brought with them a motorized vehicle (a "rover") that enabled the astronauts to travel much further on the surface and to collect much greater quantities of rock and soil. Here are some of the highlights:

- The *Apollo 15* crew of David Scott, Al Worden and Jim Irwin were blackballed by NASA after the flight and never went into space again. Their misdemeanor was to have sold a set of first-day-of-issue stamped envelopes that they had smuggled onboard in addition to 250 official ones. NASA confiscated the envelopes but had to return them twelve years later when Worden sued, with NASA settling before the case went to court. By this time the value of the envelopes had increased from about $8,000 each that the crew had negotiated with a German buyer to half a million dollars or more.
- Scott had shown a less commercial side of his nature when he undertook a school science experiment on the Moon's surface. In front of the TV camera he dropped a hammer and a feather showing that they reached the ground simultaneously because of the lack of air to slow the feather down.[1]
- The rover was light and in the Moon's low gravity it bounced – "a real bucking bronco" according to *Apollo 15*'s Scott. On *Apollo 16* John Young drove the rover flat out as part of a planned test drive. At one point all four wheels left the ground as he bounced over the rough surface.[2]
- The clinging lunar dust was a constant headache made worse by re-entering the Lunar Module between each of the three Moon walks on each mission. *Apollo 16*'s Charlie Duke fell down a few times and found that the easiest way to get back up in his awkward spacesuit was to roll into a small hole or crater that covered him in dust even more.
- The three missions all landed in mountainous areas, building on the navigational expertise built up in the earlier flights and producing images of exceptional scenery, as well as of geological interest.
- *Apollo 16* experienced several failures during its three-day coast to the Moon, at one point suggesting that it might have to return to Earth without going into orbit around the Moon. One of these required a software patch to be uploaded to the crew, who entered it into the computer, allowing the mission to continue.
- *Apollo 15* and *16* each left behind a small 80-pound (36-kg) satellite in orbit around the Moon that provided measurements of the magnetic and electrical fields in space and that could be tracked by Earth-bound radars to analyze the Moon's gravity field. The *Apollo 15* probe was tracked in orbit for 17 months and is assumed to have crashed into the surface sometime thereafter. The *Apollo 16* probe lasted only one month in orbit before crashing. It had been released in a lower orbit than *Apollo 15* had.
- One of the three parachutes failed to open properly, making the *Apollo 15* splashdown harder than normal – 36 miles per hour (58 kmph) instead of 22 miles per hour (35 kmph) – but the astronauts were unharmed.
- The three-week quarantine on return to Earth was scrapped for these missions. The Moon is sterile and free of alien bugs.

---

[1] Visible on YouTube by searching for "Apollo 15 hammer-feather drop"
[2] Visible on YouTube by searching for "Apollo 16 lunar rover grand prix"

- The last of the missions, *Apollo 17*, was the only one launched at night, making it a particularly spectacular occasion, turning night into day for the spectators all along the eastern Florida coast.
- *Apollo 17* astronaut Harrison (Jack) Schmitt was the only trained geologist among the 12 men to have walked on the Moon. His colleague Gene Cernan gained the title "the last man on the Moon" when he re-entered the Lunar Module at 23:41 on December 13, 1972, Houston time (see Figs. 6.2 and 6.3).[3]

Fig. 6.3. *Apollo 17* astronaut and geologist Harrison Schmitt is dwarfed by a large rock during the third Moon walk of the last Apollo mission on December 13, 1972. Years later *Apollo 12* astronaut Alan Bean painted a picture of the scene and inscribed the name of the other *Apollo 17* astronaut Gene Cernan's daughter Tracy into the painting, where the scuffmark is on the lower left of the rock. Reality imitated art when "Tracy's Rock" caught on as the name of the boulder in this impressive photo. (Illustration courtesy of NASA.)

References

1. Wolfe, T., *The Right Stuff*, Bantam Books (New York NY), 1980, pp. 138-145.

---

[3] 05:41 GMT on December 14, 1972.

7

Legacy of Apollo

Political

The May 1961 decision to send men to the Moon was taken on Cold War strategic political grounds. The main motivation was not to advance science or to explore the universe. President Kennedy was blunt about this in a discussion with the head of NASA, James Webb, a year and a half later[1]: "The Soviet Union has made this a test of the system. Everything that [NASA] do ought to really be tied into getting onto the Moon ahead of the Russians. This is the way to prove your pre-eminence." Other space programs were of little or no interest to Kennedy because they didn't address this political objective of global pre-eminence: "All these programs which contribute to the lunar program [that] *are essential* to the success of the lunar program, are justified. Those that are not essential to the lunar program, that help contribute over a broad spectrum to our preeminence in space, are secondary."

Apollo did what Kennedy had demanded: putting an American on the Moon and safely returning him before the end of the 1960s. As space historian Roger Launius put it 30 years later "It was an endeavor that demonstrated both the technological and economic virtuosity of the United States and established national preeminence over rival nations."

In Chapter 4 we noted the impact of the photos of Aldrin and Armstrong on the Moon, plus the fact that their steps were followed as they happened by TV viewers around the world. The Soviets were notorious for being secretive about their space missions until they had actually happened, which people began to recognize as the

---

[1] Audio recording of this meeting is available at: https://www.jfklibrary.org/Asset-Viewer/QqNoKVCaXkOHTKx6DRIvvw.aspx and a transcript of the meeting is at: https://history.nasa.gov/JFK-Webbconv/pages/transcript.pdf accessed October 9, 2018.

1960s progressed. American openness about the Apollo missions contrasted with Soviet furtiveness, and that in itself was an important message that had a subtle but wide political impact. In any interaction with the Soviet Union you worried that not all was being revealed.

In some ways the important thing is that the Soviet Union *didn't* get men on the Moon first. The Soviet economic system was much weaker and more fragile than that of the United States, so a space victory would have compensated to some extent for that weakness. U. S. economic leadership meant that victory in the race to the Moon was the icing on the cake, illustrating what people around the world knew in their gut – that the Coca Cola signs in every city, town and village signaled the superior life style of the American system. Apollo gave them a feel-good technology factor to go with that.

It's difficult to point to specific international political gains that Apollo achieved. A month after the October 1962 Cuban missile crisis that almost triggered World War III President Kennedy agreed with NASA chief James Webb that the United States "might not have been as successful on Cuba if we hadn't flown John Glenn and demonstrated we had a real overall technical capability here." American leadership in space was therefore a demonstration that it could overcome almost any technical challenge it set itself, which gave U. S. diplomacy extra credibility and punch.

Within the United States, the political legacy of Apollo was mixed. The aerospace industry benefited from the huge investments made by NASA in several states – California, Florida, Texas and a few others, and politicians in those states continue to support a strong U. S. space program. NASA also helped to set up laboratories and institutes dedicated to space research at universities across the United States. This investment has had long-term impact, since most of those labs and institutes still exist and provide support to the space program in their state. But inside NASA itself many officials failed to appreciate that Apollo had not been a normal situation and would not be repeated. It was a one-off occurrence triggered by Cold War politics. The political will would never exist again to devote 4% of the federal budget to a space program.

Social

There is much anecdotal evidence that the Moon landings helped persuade young people to follow a career in science and technology. There are of course many examples of people who became interested in astronomy or space engineering, but the example of Nobel Prize winning geneticist Paul Nurse is perhaps more noteworthy. He described in a 2014 TV interview how the excitement of the Apollo missions persuaded him to take up biology. Amazon founder Jeff Bezos said in

2016, "When I was five years old I watched Neil Armstrong step onto the Moon and it imprinted me with a passion for science and exploration" [1].

Another group of people who often point to the impact of Apollo missions on their careers is the "green" community. In this case the legacy springs not so much from the actual Moon landings but from the images of Earth captured by the astronauts. The impact began with the images taken at Christmas time in 1968 by the *Apollo 8* astronauts (see Fig. 7.1). They were the first human beings to travel far enough away from Earth to see its full disk. The images they took of Earth hanging above the horizon of the desolate Moon struck a chord with many people.

Fig. 7.1. The iconic *Apollo 8* image of Earth rising above the Moon (Earthrise) taken on Christmas Eve 1968. The contrast between the forbidding crater-strewn Moon and the beautiful and welcoming blue and white Earth emphasized the unique position of our planet as the only habitable world in an otherwise stark and inhospitable universe. On Earth a quarter of a million miles away the sunset terminator crosses Africa. The South Pole is in the white area near the left end of the terminator. North and South America are under the clouds. (Illustration courtesy of NASA.)

Later Apollo missions took many more such images, and they are still a mandatory element of every robotic lunar probe. Earthrise images from Japan's *Kaguya* (2007), China's *Longjiang-2* (2018) and India's *Chandrayaan-1* (2009) robotic probes have all been proudly published by those countries to demonstrate that their space missions are mindful of the fragile Earth. NASA, too, has not forgotten how to take these images, and one of the best of the recent crop was taken by the *Lunar Reconnaissance Orbiter* in 2015 – see Fig. 7.2.

Fig. 7.2. Earthrise image captured by NASA's *Lunar Reconnaissance Orbiter* on December 8, 2015. The dark lunar surface enhances the blue marble appearance of Earth. Earth is shown with north at the top. Africa is visible on the right, and much of South America more dimly on the left. (Illustration courtesy of NASA.)

Technical/Managerial

The management of the Apollo program was recognized at the time as setting the standard for how to control a complex high-risk endeavor. The first step was rigorous identification of every activity needed to buy-in, assemble, test and deliver every item from nuts and bolts to pieces of equipment to whole rocket engines. Engineers then estimated the cost and time for each of these myriad activities and reported that to a central management organization. This information was put together by industry and NASA management to provide a reliable picture of what would be done when and at what cost. It meant that as those myriad activities actually happened the time and cost could be compared with the earlier estimate and the overall budget and schedule could be adjusted.

This sort of "list making" or spread-sheeting seems like common sense today, but in the 1960s it was a novelty – and not always welcomed by the engineers who previously would tell their boss when they had finished a task and not before. The military were just as influential in applying this technique as NASA, and between them they helped change the culture of customers such as government agencies and of their suppliers.

An important legacy of Apollo is to have made people aware that sound management is crucial. As mentioned earlier, not the least of the features of NASA's management was making compromises when probabilities favored it, thereby avoiding "gold plating" and over-engineering, which stretch out the time and money required without significantly improving the outcome.

The importance of sound management to the Apollo program is reflected in the saying "We put men on the Moon so why can't we [insert desired outcome]". The desired outcome might be to make the trains run on time or to collect the garbage regularly or to fill in potholes in the highway or to cure cancer. If the government could manage the most difficult task ever undertaken (men on the Moon) surely it can do these more down-to-Earth tasks. So an important legacy of Apollo is to have made people aware that sound management is crucial in any task, and especially those that involve many different organizations and technologies.

Apollo required special technology to succeed. The Saturn V rocket was five times more powerful than the rockets needed for any other use of space – satellites for weather forecasting, broadcasting, navigation or environmental monitoring, for example. Robotic missions to the planets, asteroids and comets can all get by with smaller rockets, and sending humans into orbit around the Earth can, too. So the legacy of the Saturn V rocket has been limited.

At the nuts and bolts level, much that was learned on Apollo has proved useful. The engine in the Lunar Module that could be throttled up and down, giving it helicopter-like ability, has been copied many times since. The fine control it offered is not so much for landing on moons and planets but for placing satellites in orbit around Earth with precision – perhaps dropping off one satellite here, then moving to a slightly different orbit to drop off another there. And there are many, many more examples of Apollo technology being used in space today, usually with improvements to take advantage of new technologies such as miniaturized electronics or digital cameras.

However, down-to-Earth uses of Apollo technologies are less obvious. It was hoped at the time that things like the hydrogen fuel cells in the Apollo CSM spacecraft would be employed on Earth to generate electricity. But these and other spinoffs have not materialized to any great extent.

Scientific

Although President Kennedy said that exploration of the Moon was a secondary priority, the scientific outputs of Apollo have in fact been substantial. Taking photos of the hidden (from us) far side of the Moon was of course interesting, as was learning the fine detail of the Moon's gravity field. But these were done by robotic probes – mainly the five lunar orbiter satellites that mapped Apollo's potential landing sites. Where Apollo really hit the science jackpot was in the astronauts bringing back samples of Moon rock and dust.

There are two basic advantages of bringing samples back to Earth as opposed to analyzing them on the Moon. First you can use a much wider range of laboratory instruments to analyze the samples here on Earth compared to the miniaturized instruments you can take to the Moon, whether on a robotic probe or a manned spacecraft. And secondly, when you invent a new instrument or improve the old one you can repeat the analysis of the sample in the lab, whereas the robotic probe left on the Moon is stuck with the technology it was built with.

The first thing scientists wanted to know from the samples was if Earth and the Moon were the same chemically (see Fig. 7.3). Of course the Moon lacks an atmosphere, but its dust and rocks contain the same chemical elements as Earth does, perhaps with subtle differences. Analyzing meteorites that fell to Earth alerted scientists to the fact that other places in the Solar System might be different. What the Apollo samples showed was that Earth and the Moon are almost identical, so they must have been formed together or close together within the early Solar System. That eliminated some of the other theories about how the Moon was formed, such as that it came from beyond Mars and was captured as it came by Earth.

The most popular theory now explains the origin of the Moon as being assembled from debris after the impact of a Mars-sized object with the early Earth. The material flung into space was a mix of that of Earth and the body that hit it, and the same mix was left on the surface of Earth – hence the similarity between their chemistries ([3], p. 10).

Scientists then wanted to analyze the age of the Moon dust and rocks. This is done by looking at the tiny amounts of radioactive material they contain and seeing how much of it has decayed. Before the samples were brought back, the age of the Moon was calculated by counting the number of craters on the surface – the more craters the older the surface. By analyzing the samples, scientists could

80   Legacy of Apollo

Fig. 7.3. Above, rock sample brought back by *Apollo 11* weighing 6½ ounces (180g) and almost 4 inches long by 2 inches high and an inch deep, about the size of a baseball sliced in two (9.5 × 4.5 × 3 cm to be precise). Two thin sections of its interior are shown in the lower images, one in cross nichols light (left), the other in polarized light (right). (Illustration courtesy of NASA/JSC.)

finally put accurate figures to the age and thus fine tune the crater-counting method. The key is to have samples from areas of the Moon with different crater counts, that is to say, from parts of the surface with different ages.

By knowing the relationship between crater count and age, the same analysis can then be done for Mars and other planets, moons, asteroids and comets around the Solar System. So getting Moon samples into the laboratory leads to a better understanding of all of the Solar System.

As Fig. 7.4 shows, scientists have been steadily performing new research using Apollo samples in the five decades since the samples were collected. Scientists

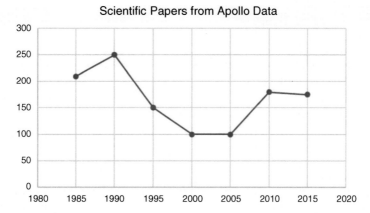

Fig. 7.4. Scientists are still consistently publishing new scientific findings based on analysis of lunar surface samples brought back by the Apollo missions. There was a burst of papers in the decade after *Apollo 11*, followed by this steady level of new findings. (Adapted from [2].)

now want to get samples from more regions of the Moon so that the crater-count analysis can be improved. A 2007 report of the National Research Council said "a vigorous near-term robotic exploration program providing global access is central to the next phase of scientific exploration of the Moon." The report recognized that the Apollo samples (and the much smaller samples from the three Soviet Luna robotic sample return missions) didn't tell the whole story and recommended that "landing sites should be selected that can fill in the gaps in diversity of lunar samples" (see Fig. 7.5) ([3], pp. 2, 5).

Scientists also want to target interesting regions such as the south pole region where there may be some deep craters that are always in darkness and may therefore contain water-ice. Some mountain peaks in the south pole region have the opposite characteristic. They are always in sunlight, on the boundary between night and day. This would be very advantageous for setting up a base because solar panels could provide electrical power all the time – and potentially a source of water not far away at the base of nearby craters.

The 2007 NRC report highlighted the region on the hidden side of the Moon near the south pole as being of special interest. This is the region to which China sent the *Chang'e 4* robotic rover, which landed in early 2019. It is special because it is the deepest large depression in the Moon's surface and so should contain materials from deeper inside the Moon than anywhere else on the surface.

In summary, the NRC report said that the Apollo samples "dramatically changed our understanding of the character and evolution of the solar system." That's quite a legacy.

82 Legacy of Apollo

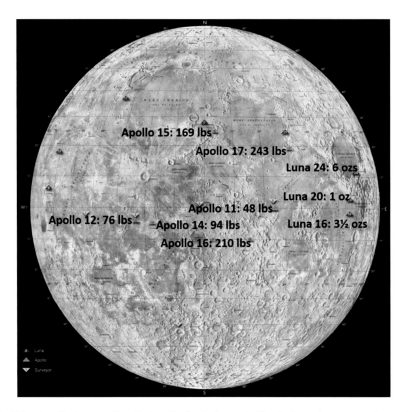

Fig. 7.5. Apollo (green triangles) and robotic Luna (red) missions that returned samples from the Moon's surface and the amount of Moon dust and rocks they collected. Other U. S. (yellow, Surveyor) and Soviet (red, Luna) probes that did *not* return samples are also shown. The sites are all on the front side (thus leaving the hidden, far side half of the Moon completely unsampled), mostly near the equator and very unevenly distributed across the surface. (Illustration courtesy of NASA and author.)

References

1. http://b-townblog.com/2015/11/19/photos-jeff-bezos-unveils-recovered-apollo-rocket-engines-at-museum-of-flight/ accessed October 9, 2018.
2. Crawford, I. A., "The Scientific Legacy of Apollo", *Astronomy and Geophysics* (Vol. 53, pp. 6.24-6.28).
3. National Research Council (NRC), *The Scientific Context for Exploration of the Moon*, National Academies Press (Washington DC), 2007.

8

The Other Competitor in the Race

Runner Up in the Race to the Moon

From 1961 to about 1965 the Soviet Union seemed to be ahead of the United States when it came to human spaceflight (see Fig. 8.1). Yuri Gagarin established this when he became the first man to orbit Earth (April 1961) followed by the first human to stay in space for 24 hours (August 1961), the first mission involving two spacecraft and two humans (August 1962), the first woman in space (see Fig. 8.6) and the first 5-day mission (June 1963), the first three-man crew in space (October 1964) and the first spacewalk (March 1965).

The Soviets also seemed to have a lead in unmanned lunar and planetary missions, starting with the first photos of the far side of the Moon in 1959 and the first flyby of Venus in 1961. This was followed by the first soft landing on the Moon's surface in February 1966 and the first satellite to orbit the Moon two months later.

The impression was of a steady increase in Soviet competence, always one or two steps ahead of the United States, that would lead eventually to a manned lunar landing.

Despite that early Soviet lead, America was first to place men on the Moon, in July 1969 with *Apollo 11*. Months and then years went by without a Soviet response. Surely if the Soviets were close on America's heels in the late 1960s they would put men on the Moon's surface if only to show that they *could* do so. Was Soviet technology simply not up to the task? The inescapable conclusion was that the United States had a significant lead over the Soviet Union when it came to Moon missions. In the end no Soviet cosmonaut ever made it to the Moon's surface. Their team in the race Did Not Finish (DNF).

The official Soviet line from 1969 on was that they had never intended to send men to the Moon. Their plan was to follow the advice of the grandfather of Soviet

Fig. 8.1. Yuri Gagarin (*left*) and Sergei Korolev in May 1961. Korolev orchestrated the catalog of Soviet space spectaculars starting with *Sputnik-1* in October 1957. The most important of these in triggering the race to the Moon was Gagarin's April 12, 1961, flight, forcing the United States to aim at a Moon landing as the only way to convincingly demonstrate superiority in space. (Illustration courtesy of the British Interplanetary Society archives. Used with permission.)

space concepts, Konstantin Tsiolkovsky, who wrote about space travel in the late 19th and early 20th centuries. He advocated the creation of space stations in orbit around Earth from which expeditions to the Moon and Mars could later be mounted. Thus the Soviets continued to send humans into Earth orbit and to increase the size of the spacecraft and the duration of their stay in space. In 1971 they launched *Salyut-1,* which they called the world's first space station. It was launched without a crew onboard, and cosmonauts were then sent to it in a separate launch. A second strand to the Soviet post-*Apollo 11* story was that robotic probes were more appropriate for returning samples from the Moon. Why place humans at risk and incur huge costs when you could achieve your objectives with inanimate machines?

This was a story devised after it became clear that America would win the race to send humans to the Moon. The real story was that the Soviets had three Moon programs underway in the late 1960s. One was to send humans around the Moon without landing, another to land men on the Moon and a third to land a robotic vehicle on the surface, pick up samples of lunar rock and return them to Earth.

The first two ended up being canceled in the usual deniable Soviet way. Their funding gradually dried up, but there was no concrete cancelation decision – a cancelation would imply that they had existed – and the Soviet story was that there had never been a manned lunar program. Thus, only the third of these was actually completed, returning a total of 326 grams (11½ ounces) in three separate missions by *Luna 16* in 1970 (101 grams, 3½ ounces), *Luna 20* in 1972 (55 grams, 2 ounces) and *Luna 24* in 1976 (170 grams, 6 ounces) – all landing near the eastern rim of the Moon (see Fig. 7.3). The automated return of samples from the Moon was a major achievement by the Soviets but a big letdown having lost the main race to put the first humans on the Moon.

The real story of the Soviet attempts to win the race to the Moon is one of high level management incompetence, which has only emerged since the end of the Cold War. That is the story that will unfold in this chapter.

Soviet rocket development during the 1950s and 1960s was driven by the need for long-range missiles that could carry nuclear weapons over intercontinental distances. Immediately after World War II, Soviet rockets were copies of the German V2 that had been used to bomb London and Antwerp. Large quantities of rocket parts, engineering equipment and more than 100 German engineers were moved from Germany to the Soviet Union to help make that happen. Note that the U. S. Army was pursuing a similar line of development using Wernher von Braun and other senior German rocket engineers who had surrendered to the U. S. Army in 1945.

One Soviet team led by Sergei Korolev (see Fig. 8.2) was soon producing rockets that significantly exceeded the performance of the V2. Acting much as an entrepreneur would in the capitalist west, Korolev persuaded both the military and the politicians on the merits of his missile concept while simultaneously managing the engineering team that was pushing the technical boundaries. One of the more visible signs of his creative management skills was a committee he formed of the other leading engineering teams involved in rocket development – the Committee of Chief Designers (see Fig. 8.3). This committee thrashed out what the engineers wanted to see happen. As in any political system two or more competing ideas mean that nothing will happen, since the bureaucrats don't want to be

86   The Other Competitor in the Race

Fig. 8.2. Sergei Korolev in 1953. His genius was to find ways to finesse the Soviet system so that the communist party chiefs, the civil servants, the military, the rocket engineers and the scientists supported his plans for rockets capable of sending humans into space. He is often presented as a brilliant rocket designer, but that accolade really belonged to others such as Glushko. Korolev was above all a master manager with a vision of space travel – much like von Braun in the United States. (Illustration courtesy of Asif Siddiqi. Used with permission.)

responsible for choosing between them. By jointly promoting a single concept the committee made it easier for the officials who held the purse strings to give their agreement.

Fig. 8.3. The Council of Chief Designers in 1959. From the left: Aleksey Bogomolov (telemetry), Mikhail Ryazanskiy (radio control systems), Nikolay Pilyugin (autonomous guidance systems), Sergei Korolev (chairman and overall design), Valentin Glushko (rocket engines), Vladimir Barmin (launch facilities), Viktor Kuznetsov (gyroscopes). The six on the right had been informally coordinating since 1946 and formed the council in late 1947. The council was separate from the official Soviet institutes and design bureaus, thus bypassing much bureaucracy and institutional bias. Because of its effectiveness the council eventually assumed engineering control over much of the early Soviet space program. (Illustration by Asif Siddiqi. Used with permission.)

The second most influential of the chief designers was Valentin Glushko, whose specialty was rocket engine development. In later years disagreements between Korolev and Glushko would cause enormous harm to the Soviet space program, but initially their collaboration was strengthened by the fact that they had both been imprisoned in the infamous Soviet gulag system. Korolev had been sentenced in 1938 on grounds of being a member of an anti-Soviet organization, enduring torture by the KGB secret police that caused permanent damage to his health and would later contribute to his untimely death. Glushko had been arrested a few months earlier on various grounds, including maintaining connections with enemies of the people. The requirements of World War II allowed Korolev to be engaged in rocket development from late 1940 sponsored by the famous aircraft designer Andrey Tupolev (and approved by the Soviet dictator Stalin) but still officially a prisoner. Glushko also was engaged in rocket development during the war while still officially a prisoner, and in fact Korolev was transferred to his unit in late 1942.

88 The Other Competitor in the Race

Having absorbed the lessons of the German V2 and their own separate developments, by 1957 Korolev and his collaborators had developed the R-7 rocket that flew 3,700 miles (6,000 km) in its August 21, 1957, test-flight – the world's first intercontinental ballistic missile (see Fig. 8.4).

Fig. 8.4. Main picture: December 15, 2015, a Soyuz TMA-19M rocket carries Tim Kopra (NASA), Yuri Malenchenko (Russia) and Tim Peake (European Space Agency) from Baikonur to the International Space Station. (Illustration courtesy of NASA/Joel Kowsky.) Top right inset: April 12, 1961, Vostok-1 carries Yuri Gagarin into space. Both rockets are similar and derive from the R-7 ICBM that first flew successfully in 1957. (Illustration courtesy of ESA.) Lower left inset: The four separate engines in the first stage of the Soyuz are strapped in a ring around the single-engine second stage, with each engine containing four main combustion chambers (in red) and two smaller fine-adjustment chambers. The 60-year successful history of this design illustrates the effectiveness of clustering rocket engines and combustion chambers as opposed to building a single giant one. (Illustration courtesy of ESA/Manuel Pedoussaut, 2017.)

Since an early age Korolev had been interested in using rockets to explore space. He had been inspired by the writings of Konstantin Tsiolkovsky, who worked out much of the theory behind rocket performance as well as describing how rockets would enable people to escape Earth's gravity, go into orbit around Earth and then reach the Moon and Mars. Korolev made sure that the rocket he developed for the military would be powerful enough to also place objects into orbit. He lobbied the Soviet Premier Nikita Khrushchev (Stalin's successor) to fund the development of a bare bones satellite and to allow it to be launched on a slightly modified RV-7 rocket. The result was the world's first artificial satellite, *Sputnik-1*, launched on October 4, 1957.

Korolev continued the theme of dual-use technology[1] with his next series of satellites. He developed a satellite that could be brought back to Earth – unlike *Sputnik*, which burned up in the atmosphere when its orbit decayed after three months. The military were interested in a returning vehicle because it could bring back film containing images of Earth seen from space. The film would then be developed and used for military and intelligence purposes.[2] Korolev designed the satellite with a pressurized capsule that retained a breathable atmosphere while in space, unlike equivalent American surveillance satellites of that era, which allowed the atmosphere to escape, thus requiring the camera equipment to function in the vacuum of space.

The Soviet satellite was of necessity more than twice as heavy as its American counterpart to provide the rigidity to retain the air. This meant that it needed a much larger (and thus costly) rocket to place it in orbit but still within the capability of the R-7 rocket. The advantage of retaining the atmosphere was that the cameras and other equipment that worked on ground or in aircraft were more likely to work in the satellite. The Americans, for example, had to develop a new form of film that would work in a vacuum, a step that the Soviets avoided by their use of a pressurized capsule.

Korolev's satellite could also be used to bring animals and even humans back from space thanks to its pressurized cabin. Having shown that it worked with dogs, Korolev got the go ahead to send the first man into space, Yuri Gagarin, in April 1961. We have seen in Chapter 1 how this achievement triggered a major reaction in the United States in the form of President Kennedy's commitment a month later to send men to the Moon "before the decade was out."

The enormous prestige of the *Sputnik* and Gagarin flights strengthened Korolev's position in the Soviet system. Premier Khruschev privately referred to him as "his magician" because of his ability to conjure up a high profile space spectacular on demand. However, there was never an official policy to explore space or to send humans into orbit. The Soviet successes came about because a group of engineers led by Korolev were fascinated by space travel and seized the opportunity of the military's need for rockets to advance their ideas.

The Soviet Union was a totalitarian state, which implies that the dictator could decide what needed to be done and make it happen without debate or consultation.

---

[1] Dual use = of interest for both military and civilian use

[2] Note that camera film had to be processed chemically in order to see the images until the emergence of digital cameras in the 1970s that produce images that could be sent by radio link to ground and which didn't "run out of" film. Russia continued to use "wet film" cameras that have to be returned to Earth to be processed well into the 21st century, when most other countries had already switched to digital technology.

This was indeed true, but it omits the role of junior officials devising new ideas and "selling" them to key senior officials so that they came to the attention of the dictator in a positive light. Under Stalin (who died in 1953) this process was potentially fatal. In practice it was even worse than that because within the top tier of communist party officials the missile program was controlled by secret Police Chief Lavrentiy Beria, along with the nuclear weapons program. Coming to the attention of Stalin or Beria could result in your projects being approved, but could just as easily result in you being killed (along with your family and friends) on nothing more than a whim.

After Stalin's death (and shortly thereafter that of Beria's) the worst excesses of his era ended so that the killings ceased. Nevertheless the system still required that the top leadership gave their go-ahead to new projects such as the space program.

The *Sputnik* flight and that of Yuri Gagarin brought the Soviet Union enormous prestige – and rightly so! Here was a country that 20 years earlier had been ravaged by Germany. All of the country to the west of Moscow had been flattened. For those living in countries not subjected to the trauma of such a war it is hard to imagine the destruction involved. All industry in the German-occupied areas was destroyed – except for that physically transported by the Soviets to the east. The Soviet Union ended up creating new cities to the east of the Ural Mountains, which Stalin decided would be far enough from the advancing German armies to be safe. One of the first discoveries of America's spy satellites in the 1960s was the existence of more than 50 of these Soviet cities previously unknown to the West.

This war-ravaged country managed a mere 12 years after the end of World War II to place the first satellite in orbit. Then four years later they placed in orbit Yuri Gagarin, who as a child had been forced by the German advance to flee his home. Not by chance, Gagarin was handsome, charming, articulate and working class, and he became a symbol of a Soviet Union that wished to portray itself as technologically and socially advanced (see Fig. 8.5). The Apollo astronauts later became ambassadors for the United States, but few if any of them ever touched the heart of the common man quite as much as Gagarin did.

By enhancing the prestige of the Soviet Union around the world, Korolev's space achievements strengthened the position of Khrushchev at home from 1957 to 1961. But in late 1962 Khrushchev's position was seriously weakened. In what became known as the Cuban Missile Crisis, the threat of U. S. military action led to the Soviet Union agreeing to withdraw missiles it had surreptitiously introduced into Cuba. In return America agreed (secretly) to withdraw missiles from Turkey and not to invade Cuba, but the outcome was widely seen around the world and within the Soviet leadership as the Soviet Union backing down when threatened by the United States.

Unable to get short- or medium-range missiles close enough to the United States to be effective, Khrushchev and the Soviet military decided that they needed

The Other Competitor in the Race   91

Fig. 8.5. Yuri Gagarin (left) with Soviet Premier Nikita Khrushchev on the balcony of the Kremlin in Red Square, Moscow, October 1961, illustrating how space and politics have gone hand in hand since day one. Gagarin's charm and good looks enhanced his ability to play the role of communism's ambassador to the world – in communist and non-communist countries alike. The soft power benefits of Gagarin's flight outweighed for a while the fact that the launcher was not suitable to be an operational military missile. (Illustration courtesy of British Interplanetary Society archives. Used with permission.)

to develop and deploy long-range and submarine-based missiles that could strike the United States without the need for a forward base such as Cuba. Korolev's R-7 rocket had the range to reach the United States, but its design made it vulnerable.

Korolev had designed a rocket using fuel – kerosene plus liquid oxygen – that produced a lot of power. The kerosene was no problem, but the liquid oxygen (which, as you know, is a gas at normal temperatures) required refrigeration equipment to cool it to –297 °F (–183° C). A missile had to be able to launch at a moment's notice, so the R-7 required a crew of operators on site continuously topping up the liquid oxygen which steadily boiled off. Plus, complex and expensive refrigeration equipment was needed at each missile site. The cost of this manpower-intensive missile appalled Khrushchev, with the result that his admiration for, and sponsorship of, Korolev fell dramatically. Fifty missile sites were to have been built for this rocket, but in the end only three were.

There were alternatives that were perhaps not quite as powerful but didn't require the hugely expensive and cumbersome refrigeration exercise. Chief Designer Glushko came forward with missile designs involving fuels that were stored at room temperatures – for example using nitrogen tetroxide instead of liquid oxygen. Such fuels were poisonous and corrosive, but of course any rocket fuel is dangerous. These fuels could be stored without the need for expensive cooling equipment and could be loaded into the missile and then left for months without any attention. Korolev argued vigorously and then vehemently that his cryogenic (low temperature) fuel combination was the best way forward because it produced a greater amount of thrust per pound/kg of fuel than the alternatives, and he refused to compromise. In the end Glushko refused to work with him any longer, and explicitly refused to undertake development of the large engines required for a Moon landing mission.

Khrushchev realized that Korolev was pursuing an agenda of his own – pushing for space exploration at the expense of Soviet military priorities. He therefore encouraged other industry groups to take on the task of developing long-range missiles. One immediate alternative was Glushko together with a supplier he recommended, Mikhail Yangel, who previously worked with Korolev. Yangel was given the green light in August 1958 to develop the RS-16 long-range missile with Glushko supplying the engines for both of its two stages. Yangel went on to develop many of the Soviet Union's most successful missiles.

In 1960 another alternative to Korolev emerged in the form of Vladimir Chelomey, who was given the go-ahead to develop a long-range missile he had proposed, the UR-200 that used storable propellants.[3] Chelomey had been developing short-range missiles for the Soviet navy, and the UR-200 was a big step up in scale for his team. Chelomey's success in persuading Khrushchev to approve the UR-200 was helped by the fact that the latter's son, Sergei Khrushchev, was one of his employees. Although the UR-200 was later canceled, Chelomey went on to develop the larger UR-500 rocket, which eventually became the Proton rocket still in use in 2019 as the largest Soviet space launch vehicle.

Despite the success of Yangel/Glushko and Chelomey in winning contracts to develop long-range missiles, Korolev also won approval for the development of a new missile, the R-9, which used liquid oxygen. One of his arguments for winning this work was that the U. S. military was developing the Titan-1 missile that involved similar technologies. Note that the Titan-2 rocket used to launch NASA's two-man Gemini capsules in 1965-66 used storable propellants in place of the liquid oxygen/kerosene of the Titan-1.

---

[3] Both Chelomey's design and that of Yangel and Glushko used unsymmetrical dimethylhydrazine as the fuel. Chelomey used nitrogen tetroxide as the oxidizer, while that of Yangel/Glushko used a nitric acid/nitrogen tetroxide mix. All of these chemicals are toxic and corrosive.

Korolev's team was also kept busy with his ongoing surveillance satellite work and his ability to produce a world-beating space spectacular to order – see the list at the start of this chapter – at least until 1966.

To the outside world the Soviets appeared to be systematically developing bigger and better manned space vehicles. And indeed Korolev was designing the Soyuz spacecraft that could carry three people, could maneuver in orbit and could dock with other spacecraft, but it was poorly funded and increasingly delayed. The spectaculars of 1962-1964 were based on slight variants of the Vostok capsule that carried Gagarin aloft in 1961. Three cosmonauts squeezed (unsafely, many would say) into a slightly modified Vostok capsule in 1964, rebranded as a *Voskhod-1*. The three cosmonauts' seats inside Voskhod were only fitted with difficulty, making the crew crane their necks to see the instruments. As there was no space to equip the capsule with a means of escape, retrorockets were fitted to slow it as it returned to Earth, enabling the cosmonauts to remain on board for the entire mission instead of having to eject and use individual parachutes.

In March 1965 an even more unsafe variant of the Vostok, called *Voskhod-2,* was used for the first "spacewalk" by Alexey Leonov (see Fig. 8.6).

The perception in the West was that the Soviets were systematically and steadily developing new space vehicles that would inexorably lead to a mission to the Moon. The reality was that each Soviet mission was a one-off project, approved with minimal lead time aimed at providing a news headline as requested by the political leadership, and timed to upstage NASA's missions.

Mostly the political motivation was to do with important meetings of Soviet or world communism, but one was part of a major Soviet foreign policy initiative – the erection of the Berlin Wall. A month before the Soviets began building the wall, Khrushchev informed Korolev that the launch of *Vostok-2* should take place within three weeks, not later than August 10, 1961. *Vostok-2* was in fact launched on August 6, carrying the second man to orbit Earth, German Titov, who spent a day in space and completed 17 orbits around Earth (Gagarin completed just one orbit). Presented as usual to the world as an unalloyed success, Titov's mission was fraught and hazardous. He suffered from severe motion sickness (as did many later astronauts) for the first twelve orbits, which prevented him from undertaking many of his assigned tasks. The re-entry into Earth's atmosphere, like that of Gagarin four months earlier, was nearly a disaster because the instrument section failed to separate fully from the spherical descent section. Fortunately the instrument section eventually burned up in the heat of re-entry (as had also happened to Gagarin), allowing the capsule to stabilize. Titov then parachuted out of the capsule as planned.[4] He remains the youngest person to orbit Earth, aged 25 when he flew *Vostok-2*.

---

[4] The Soviets publicly stated that Gagarin and Titov stayed in their capsules all the way to the ground so that they could unambiguously claim the first launch of a human from Earth into space and his safe return to Earth.

94 The Other Competitor in the Race

Fig. 8.6. The first woman in space, Valenitina Tereshkova, and the first human to perform a spacewalk, Alexey Leonov, reminisce in Moscow on the occasion of the 50th anniversary of Yuri Gagarin's flight, April 12, 2011. Tereshkova in *Vostok 6* and fellow cosmonaut Valeriy Bykovskiy in *Vostok 5* were in space at the same time in June 1963 but in different orbits that prevented them seeing each other – a blatant publicity stunt of no significant technical value. Tereshkova was unfairly criticized after the flight for poor performance by Korolev and other officials in what seems to have been a glaring case of sexism (much worse performance by male cosmonauts went unremarked). After Leonov's hazardous March 18, 1965, spacewalk (see text) the United States took the lead in human spaceflight with the deployment of the Gemini two-man capsules and then the Saturn V/Apollo Moon ship. (Illustration courtesy of NASA.)

The following week, on August 13, 1961, the erection of the Berlin Wall began – dividing Berlin into East and West, and extended in the form of barbed wire-topped walls and fences to divide West and East Germany until November 1989. The event was seen in the West as proof of the failure of the communist system – having to prevent its own citizens from fleeing by fencing them in. Meanwhile Titov and Gagarin were sent on good will tours around the world, visiting 11 countries in 1961 and many more the following year. Although the Berlin Wall exposed the dark side of Soviet society, the success of its space program continued to give an impression of technological excellence. However, the one-off character of the spectacular space missions had the effect of delaying systematic and steady Soviet progress by diverting funds away from the development of the Soyuz capsule and other important new systems.

Leonov's spacewalk in *Voskhod-2* in March 1965 proved to be the last gasp of this stopgap approach. An unmanned version of the craft was launched on February 22 and worked well until a spurious signal from ground control caused it to descend early, resulting in its destruction during the uncontrolled re-entry. Without waiting for further flight tests, Alexei Leonov and Pavel Belyayev were launched on a day-long mission on March 18, 1965, just five days ahead of the first American Gemini capsule, and less than three months ahead of the second Gemini mission, during which a U. S. astronaut would perform a spacewalk, or extra-vehicular activity (EVA) in NASA-speak.

An hour and a half after takeoff, the cosmonauts commanded the concertina-style airlock that had been added to what was otherwise the standard Vostok capsule (as used by Gagarin, Titov, etc.) to expand to its full size. Leonov entered the airlock, then opened the door and pushed himself into space. He took the lens cap off the external camera, which sent images back to Earth, and tried unsuccessfully to take photos of the spacecraft using his own camera. He then tried to re-enter the *Voskhod-2* airlock only to find that his spacesuit had ballooned so that he could not enter feet-first. He had to partially deflate the spacesuit, enter the airlock head first and then perform a head-over-heels maneuver inside the airlock while still wearing the spacesuit. His heart rate hit 143 beats per minute, his body temperature exceeded 100° F (38° C) and he was drenched in sweat. The internal diameter of the airlock was not designed for a somersaulting cosmonaut! The whole episode lasted 23 minutes, 12 minutes of which Leonov was actually in space.

Needless to say the mission was portrayed to the world as a brilliant success and proof once again that the Soviet Union led the United States in space technology.

The problems for Leonov and Belyayev didn't end when Leonov was back inside the capsule. The descent to Earth after a day in orbit was complicated by various technical glitches, one of which resulted in the cosmonauts experiencing 10 g[5] forces for a short period, bursting blood vessels in both men's eyes. The craft was 250 miles (400 km) off course, ending up in the middle of a Siberian forest in deep snow. The cosmonauts had to spend two nights in the capsule before a suitably equipped rescue team arrived. The nearest point where a helicopter could land was more than a mile away.

By comparison, NASA's Gemini spacecraft was designed to allow astronauts to perform an EVA without the need for acrobatic exertions. Gemini could also maneuver in space, could physically connect to another Gemini vehicle and could support a two-week-long mission. The fact that Voskhod/Vostok couldn't maneuver, couldn't dock with another capsule and was limited to five-day missions was not known in the West until some years later.

---

[5] Ten times the weight of gravity – a 150-lb man feels as if he weighs 1,500 lbs ($\frac{2}{3}$ of a ton).

96   The Other Competitor in the Race

Voskhod-2 achieved its objective of convincing the world that the Soviets still led the United States in human spaceflight in March 1965. But it was downhill after that. The eight Soviet manned space missions up to that point had all used variants of the Vostok spacecraft that had carried Gagarin into orbit four years earlier. As NASA's Gemini spacecraft began its series of missions (March 1965) the Soviets had no tricks left up their sleeves with which to outshine them. The years 1966 and 1967 were disastrous years for Soviet human spaceflight, with no missions launched. Meanwhile NASA's Gemini program was an astounding success, demonstrating most of the suite of techniques needed for a Moon landing mission, including rendezvous and docking of two spacecraft in space, transfer of crew between two docked spacecraft, and missions lasting more than the ten days needed for a Moon landing (see Fig. 8.7).

Fig. 8.7. The United States achieves the first fully successful rendezvous of two spacecraft in orbit on Dec. 15, 1965. *Gemini 7*, with Frank Borman and Jim Lovell onboard, is 37 feet (11 m) from Wally Schirra and Tom Stafford in *Gemini 6*. The spacecraft were not equipped to dock with each other, but Schirra maneuvered *Gemini 6* to within 1 foot (30 cm) of its sister craft. (Illustration courtesy of NASA.)

The Soyuz capsule was to have been the equivalent of the main American Apollo module – a three-man spacecraft that had the versatility to fulfill a mission to the Moon as well as Earth-orbiting missions. It was specifically designed to carry cosmonauts into orbit around the Moon and with the addition of a lander (the equivalent of the Apollo Lunar Module) could have been part of a Moon-landing mission. Approved for development in late 1963, development of the Soyuz fell behind schedule, because Korolev's team was diverted onto the Voskhod missions and also because its funding was stopped in August 1964, when Korolev's rival, Chelomey, persuaded the Soviet communist party and government to use his technology for the "around the Moon" mission (more about this below).

This is just one example of the torturous decision-making arrangements in the Soviet Union. One space historian describes "the mind boggling confusion of hierarchy that was the Soviet space program in the 1960s" but notes that this was "tempered by institutional loopholes that allowed design bureaus and chief designers to push their programs through informal channels" ([1], p. 286).

Korolev decided to re-badge Soyuz as an Earth-orbiting spacecraft that, like NASA's Gemini, could be used to prepare for a lunar mission but not actually undertaking one. By this sleight of hand Korolev obtained approval in February 1965 for development of Soyuz to continue. We will come back to this story below.

We have seen in Chapter 2 how the key technological development for the U. S. Moon landing was the giant Saturn V rocket. The Soviets, too, recognized that a manned lunar landing would require a rocket much more powerful than the ones that placed cosmonauts into Earth orbit. Both countries started out planning to launch at least two giant rockets and assemble their payloads into a Moon "ship" in Earth orbit, then send the assembled craft to the Moon for the landing and return.

As we have seen, U. S. planners found a way to do the job with just one launch of a giant rocket carrying a complex spacecraft. That single-launcher scheme called for placing the spacecraft into orbit around the Moon, then sending down a module to the surface that could return to the mother ship in orbit above, and from there return to Earth. The Soviets, especially Korolev's team, carried out similar studies and followed the U. S. events closely. (The reverse of course was not possible, since the Soviets released very little information.)

Korolev's N1 rocket was eventually selected as the launcher for the Soviet manned lunar landing program, but without the whole-hearted backing that the Saturn V got in the United States. On the face of it, the N1 seemed to get the clear go-ahead as the Soviet Union's future heavy-lift rocket in a joint decree of the Soviet Union Council of Ministers and the Central Committee of the communist party issued on September 24, 1962 ([1], p. 331). That called for the N1 to place a payload of 75 tons[6] in Earth orbit with a first launch in 1965. It identified the possibility of increasing the payload capability later using upper stages with more advanced fuel technology, such as liquid hydrogen. The main mission identified for the N1 was for a military heavy missile called an orbital bombardment system, and nothing was said about a Moon-landing mission. The decree said that the goal was to ensure the leading position of the Soviet Union in the exploration of space, the sort of phrase that politicians love because it is so general that almost any future outcome can be said to have been met.

The development did not go smoothly! While the United States assumed that the Soviets were focused on competing with President Kennedy's challenge to send humans to the Moon, the various factions in the Soviet space community were in fact focused on competing with each other. Korolev's engine designer from the days of Sputnik and Gagarin, Valentin Glushko, refused to work on the N1, arguing that its liquid oxygen technology was inappropriate for military launches. Korolev was forced therefore to use an engine supplied by the relatively inexperienced Kuznetsov team, and the comparatively limited performance of these engines led to a very complex design of the rocket (the first stage had 30 of these engines strapped together).

Moreover, the military was lukewarm at best about its need for such a rocket and refused to release the full amount of funding that Korolev requested.

Korolev's team was also extremely busy on other projects during this period ([1], p. 381). Missile-related projects absorbed more resources than space work, including versions of the R-9 Intercontinental Ballistic Missile (ICBM), two solid propellant ballistic missiles (RT-1 and RT-2) and the GR-1 orbital bombardment system mentioned earlier. On the space side the projects included the strategically critical *Zenit-2* reconnaissance satellite, the Vostok manned spacecraft and its variants, the *Molniya-1* communications satellite, the *Elektron* scientific satellite and robotic space probes to be sent to the Moon, Venus and Mars. There is a case to be made that the Zenit satellites were Korolev's most important contribution to history because they made possible the first Strategic Arms Limitation Treaty (SALT-1) that brought an end to the escalation of American and Soviet nuclear forces and thus helped to prevent a catastrophic nuclear war (See [2]).

Competition from other teams continued despite the September 1962 decree. For example in 1965 Glushko proposed the use of his RD-270 engine in place of Kuznetsov's in the first stage of the N1. Glushko's engine was expected to be four times more powerful than Kuznetsov's, but it was only theoretical, as its development had been slower than planned. The attraction of needing far fewer than 30

---

[6]There are at least three different weights that are pronounced "ton." For brevity, we use the word "ton" to signify a weight of 1,000 kilograms (about 2,205 lbs) instead of "tonne" or "metric ton". Note that in the United States and Canada, "ton" usually means 2,000 pounds, while in the rest of the world it usually means 2,240 pounds.

engines in the first stage was obvious, but such a radical redesign would have set the N1 development back two or three years, so in the end Glushko's proposal was rejected.

While development of the N1 limped along, the political wish to stay ahead of the Americans continued. In the early 1960s these political whims were addressed by various ad hoc and dead-end modifications to the Vostok spacecraft. By 1965 the United States had taken the lead in human spaceflight in Earth orbit, so it was clear to all parties in the Soviet Union that they would have to show leadership with a Moon mission of some kind.

In 1965 the political imperative of pipping the Americans to the Moon caused another change of emphasis. Chelomey's team was falling behind in its development of a spacecraft to send astronauts around the Moon and straight back to Earth – a circumlunar mission. Korolev tried to have the project transferred to his group, but in the end Chelomey continued with his development of the UR-500 rocket for the mission while Korolev supplied the spacecraft derived from the Soyuz he was already developing. Korolev saw this as a missed opportunity to use the same rocket for both the Moon landing and the circumlunar missions and avoid developing two completely separate rockets. Chelomey's UR-500 never made it to the Moon, but it did eventually become, and still remains, the workhorse heavy-lift launcher of the Soviet (and later Russian) space program, what is now called the Proton rocket.

Then, on January 14, 1966, Korolev died at age 59 of complications during what was expected to be routine surgery to remove a bleeding polyp in the intestine. The surgeon discovered a malignant tumor the size of his fist and removed it with great difficulty, resulting in extensive bleeding. Half an hour after the operation ended Korolev's heart stopped and couldn't be restarted.

We will never know if Korolev could have pulled off one of his "magic tricks" and beaten the United States to the Moon, but certainly without him the chances were greatly reduced.

Chelomey chose this moment to propose an alternative to the N1. His proposal, the UR-700 rocket, was more powerful than Korolev's N1 and similar in capability to NASA's Saturn V. Chelomey proposed to use the powerful engines that Glushko had proposed for the N1 a year previously, but there had been no visible progress in the development of those engines in the intervening period and that proved to be its fatal weakness. In late 1966 Chelomey's UR-700 was rejected and the N1 confirmed again as the choice for landing a man on the Moon.

The original concept for the Soviet Moon landing was to assemble the necessary spacecraft (called the L3) in Earth orbit using three N1 launches. A fourth smaller launcher would bring the cosmonauts into space, where they would link up with the assembled spacecraft. By 1965 this scheme had been replaced by a copy of America's Apollo mission – a single giant launcher takes the L3 spacecraft into orbit around the Moon, ejecting a small lander to bring one or two cosmonauts to and from the Moon's surface. This scheme involved the space vehicles performing a Lunar Orbit Rendezvous (LOR) as opposed to the Earth Orbit Rendezvous (EOR) for the three launch and assembly approach. The great advantage of LOR was that it would involve only one N1 launch instead of three, and thus should be much less expensive. The big disadvantage was that it required the N1 to be similar in performance to the American Saturn V rocket, which could lift about 130 tons into Earth orbit.

Originally proposed as capable of placing a 50-ton payload into orbit around Earth, Korolev had increased the performance to 75 tons by increasing to 30 the number of engines clustered together in the first stage. To make the LOR scheme work, he now proposed that the N1 would place 90 tons into Earth orbit, and he reduced the size and weight of the Moon-bound L3 spacecraft drastically. A consequence of that smaller spacecraft was that only one cosmonaut would travel to the Moon's surface while a second cosmonaut waited in orbit around the Moon for his return. In comparison the Apollo mission took two astronauts to the Moon's surface while a third waited in lunar orbit. The vehicle that landed on the Moon would weigh 5 tons compared to the 15 tons of the Apollo lunar lander.

Within his own team Korolev's Moon-landing mission was heavily criticized as both unrealistic and insufficient. Sending a single man to the lunar surface placed an enormous and complex burden on that person, first to navigate and land the craft, then to leave the craft and collect samples, and finally to navigate the return to the spacecraft in lunar orbit. The ability of a single person to perform these tasks had not been demonstrated and was considered impractical by many.

For most previous space missions the Soviet Union's spacecraft were heavier than the equivalent American spacecraft. The American lead in lightweight electronics and materials enabled them to design much more efficient space vehicles. Up until then, the more powerful Soviet rockets had more than compensated for their inferior spacecraft technology. Now Korolev proposed to turn that on its head and have the Soviet Union use a less powerful rocket and build a more efficient spacecraft. One of his team said this whole scheme "was on the brink of fantasy" ([3], pp. 476-477).

Nevertheless the Korolev concept went ahead, and his designers attempted to both increase the performance of the N1 rocket and to reduce the mass of the L3 spacecraft. At the time of Korolev's death, the gap between these two conflicting requirements was still wide. His motivations for proposing what seemed to be a hare-brained scheme were twofold. First it kept him in the launcher business against Soviet competitors such as Chelomey and Yankel, who had by now won almost all contracts for military missiles because they had elected to use storable propellants. Korolev stuck with the use of liquid oxygen (which required constant topping up as well as refrigeration to store it) because it had the potential to place much larger payloads into space.

Second he reasoned that the Americans might hit problems with developing the Saturn V or some other part of the Apollo program that would give Korolev time to improve the performance of the N1 rocket. He had already identified a way to add significant performance to the N1 by developing a rocket engine fueled by liquid hydrogen and liquid oxygen that could be used on the upper stages of the N1. This fuel combination was used by NASA for the second and third stages of the Saturn V and was the reason why with a smaller first stage it could loft into orbit a larger payload than the N1. The Soviets lagged behind the United States in the ability to manufacture and exploit liquid hydrogen, so this plan required time if it was to succeed.

Having ignored President Kennedy's 1961 decision to initiate the Apollo program until 1964, by 1967 the Soviet Union had three programs underway intended to beat NASA's Apollo to the Moon (see Fig. 8.8). Two of the three were intended to beat Apollo only in the sense of gaining publicity.

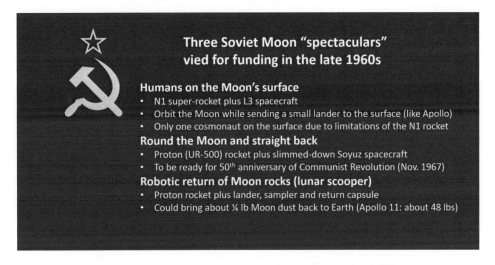

Fig. 8.8. The scattershot Soviet response to the Apollo challenge

First was Korolev's N1/L3 program to land humans on the Moon's surface in a direct race to meet the challenge posed by President Kennedy. It was recognized that this was at risk of being pipped to the post by Apollo.

Second was the separate program to send cosmonauts around the Moon but not land on it. This was to use the UR-500 (Proton) rocket being developed by Chelomey plus a stripped down variant of Korolev's nearly developed Soyuz spacecraft called the L1. By late 1965 this project was recognized as the most reliable source of a space spectacular in time for the politically important November 1967 50th anniversary of the communist revolution, and for that reason it took funding priority over the N1/L3 Moon landing project.

Third, in early 1967 the idea of a robotic lunar sample return mission was proposed and accepted. The idea was to land a probe on the Moon's surface, where it would scoop up some Moon dust and rocks and then launch itself back to Earth. This was motivated by the likelihood of losing the race to land humans on the Moon. One historian explained it as follows: "If all else failed – and Apollo was about to land on the Moon – then [they] could dispatch one of these robots to recover soil before any American astronaut. It was a pragmatic public relations exercise with important scientific payback" ([3], p. 641). Indeed we saw in Chapter 4 that one such probe was dispatched to the Moon a few days before *Apollo 11,* and crashed while attempting to land. The source of this "lunar scooper" concept was Georgiy Babakin, to whom in 1965 Korolev had handed over all work on automated deep space exploration (robotic probes to the Moon, Mars and Venus).

Let's now look at how these three attempts to outdo Apollo panned out.

The Final Lap: The Circumlunar Mission

The second of these schemes – the UR-500 (Proton)/L1 around the Moon (circumlunar) mission – was in some ways the most important in the long run. In 1965 Korolev had persuaded his superiors to use a two-man variant of his Soyuz three-man spacecraft for this mission in place of a design proposed by Chelomey. The Soyuz was Korolev's next-generation replacement for Vostok, intended as a workhorse capsule for human spaceflight. It could dock with other vehicles, change its orbit in space, undertake missions lasting up to two weeks long, and make space walks possible without the dangers and constraints faced by Leonov in the Voskhod (see earlier in this chapter). It made sense therefore to use it for the circumlunar mission, which would last five or six days.

However, the development of the Soyuz did not go well:

- The first unmanned test flight in November 1966 was intended to involve two launches, with the two Soyuz craft rendezvousing in orbit. The first spacecraft sprang a fuel leak once in orbit, and its engines therefore became unusable, so the second launch was canceled. The craft in orbit was given the anonymous *Cosmos 133* name by the Soviets in order to disguise its true mission. It went off course during re-entry, and its self-destruct mechanism blew it up.
- Two weeks later the second Soyuz (unmanned) was launched, but the rocket exploded on the launch pad. One person was killed and several severely injured, and many top officials had a narrow escape – including Korolev's successor, Vasiliy Mishin. The launch pad was totally destroyed, which delayed the overall program.
- The third Soyuz, also unmanned, was launched two months later (February 7, 1967), labeled unhelpfully *Cosmos 140.* It suffered some minor failures

that a cosmonaut aboard might have worked around, but its return to Earth was compromised by a hole appearing in the side of the craft. This would have proved fatal for any cosmonaut on board, and in the event it caused the vehicle to land way off course on an iceberg in the Aral Sea,[7] into which it eventually sank – retrieved from a depth of 10 meters with great difficulty by a combination of divers and a heavy-lift helicopter.

Even though the three unmanned Soyuz test flights had ended in failure, it was decided to make the next launch a manned one. It had been two years since the Soviets had launched a human into space, which for the political leadership looked like a sign of failure – two years during which ten hugely successful American Gemini flights carried sixteen U. S. astronauts into orbit. There was also pressure to do more than just launch an unmanned craft for the upcoming 1967 May Day – one of the biggest holidays in the communist calendar and the occasion for parades in Moscow's Red Square in front of the communist leadership. The dangers inherent in human spaceflight had been highlighted by the February 21 fire in the *Apollo 1* capsule that had killed three U. S. astronauts during ground testing. The head of Soviet cosmonaut training, Lt. General Kamanin, noted in a March 7 interview with Warsaw Radio that the next Soviet manned flight would only happen when it was assured of success (subtly implying that NASA had pressed ahead too hurriedly with its program).

Sadly, despite General Kamanin's reassuring words, the manned Soyuz-1 mission ended fatally. Launched on April 23, 1967, it carried 40-year old cosmonaut Vladimir Komarov on his second trip into space. The plan was to meet up with a second Soyuz that would be launched shortly afterwards. Once in orbit, several problems were observed, including a solar panel that had not deployed and a failed sensor needed to determine the vehicle's pointing direction. The jammed solar panel meant that electrical power would run out after about a day, while the broken sensor meant that rendezvous with another Soyuz would be impossible. The second launch was canceled and Soyuz-1 returned after just over a day in space. Skillful work by Komarov got the Soyuz into the right orientation for the return, but the parachute failed to open and the capsule hit the ground at 90 mph (140 kmph), killing him instantly. Some minutes later the remaining fuel caught fire and the capsule burned for several hours, but Komarov was already dead.

Sending a cosmonaut around the Moon before the October 1967 50th anniversary celebrations now looked very unlikely. And the chances diminished even more as the rocket to take it to the Moon started to fail. The Proton rocket had shown good reliability in its early launches – five of the first six launches were successful, which is a good record for a new rocket. The seventh Proton launch

---

[7] At that time the Aral Sea was the third or fourth largest lake in the world (Lake Victoria in Africa was about the same size); now it is less than a tenth that size due to the rivers that feed it being diverted for irrigation.

was in September 1967 carrying an unmanned L1 spacecraft (a much modified Soyuz) bound for a trip around the Moon and back. But one of the six main engines failed to ignite, and the rocket veered off course, crashing about 40 miles (65 km) from the launch pad, releasing its fuel[8] in a cloud of toxic yellow-brown gases.

Two months later, and after the fanfare of the 50th anniversary celebrations were out of the way, the next L1 circumlunar spacecraft was launched onboard a Proton rocket on November 22, 1967. This time the spacecraft crashed about 250 miles (400 km) from the launch site. One of the four engines in the second stage failed to ignite. Despite this, Mishin was hopeful that he could achieve a manned circumlunar flight before the Americans. The latter were not expected to begin manned Apollo flights until late 1968 at the earliest.

However, the failures kept coming. The March 2, 1968, launch worked fine and took the L1 payload out a quarter of a million miles to simulate a lunar mission, but the re-entry into Earth's atmosphere went askew due to a failed orientation sensor. A crew onboard would have experienced forces of up to 20 g but probably would have survived. The craft was heading for just off the west coast of Africa when it was deliberately destroyed to prevent it being "captured" by the United States.

The next launch five weeks later failed to reach orbit because the spacecraft computer detected a nonexistent breakdown and instructed the rocket to shut down. The payload recovery system worked perfectly, and the L1 capsule was recovered unharmed 325 miles (520 km) away and brought back to Moscow. More launches were planned for July, August, September and October, but the first two were abandoned due to an accident on the launch pad that killed one technician and left the fully fueled Proton rocket leaning perilously, held upright only by the emergency escape tower. It took two weeks of cautious round the clock work to remove the hazardous and volatile fuel and dismantle the rocket and its payload.

The first spacecraft to circle the Moon with live passengers was *Zond 5*, launched on September 15, 1968. The main passengers were two tortoises accompanied by various fruit fly eggs, herbs, algae, bacteria and plant cells. It successfully swung around the Moon at an altitude of 1,100 miles (1,750 km) and headed back to Earth, capturing high-quality photos of the complete Earth on route.[9] Once again, though, the orientation sensors broke down, and the orientation system itself failed, so the 2-ton probe could not be precisely controlled on its return. The best that could be done was to aim it in the direction of a fleet of back-up recovery ships in the Indian Ocean. Nearly seven days after being launched *Zond 5* splashed down in darkness 65 miles (105 km) from the nearest Soviet recovery ship. Several hours later it had been hauled on board, watched closely by an American navy ship. The tortoises survived!

---

[8] The fuel was unsymmetrical dimethyl hydrazine and the oxidizer was nitrogen tetroxide, both very poisonous to humans.

[9] Three months before the more famous *Apollo 8* photos.

Meanwhile, in the United States the first successful manned Apollo mission was launched on October 11, 1968 – *Apollo 7,* whose three-man crew orbited Earth for eleven days. Having seen the apparent success of *Zond 4* and *Zond 5,* NASA management recognized that a manned Soviet mission around the Moon was imminent, so they announced that the next Apollo mission in late December would swing around the Moon.[10] Despite this clear statement of intent by NASA, the Soviets decided to continue with three more unmanned Zond missions before attempting a manned mission in January 1969. *Zond 6* was duly launched on November 10, 1968, taking excellent photos of Earth and the Moon but crash landed on its return (due to the release of the parachute too early), killing the biological samples.

The crash of *Zond 6* prevented any further Soviet circumlunar launches before NASA's *Apollo 8* in late December.[11] And once *Apollo 8* proved to be an enormous success (technical, political and public), the Soviet manned circumlunar program was quietly ended without any official admission that it had ever existed.

The Final Lap: Robotic Sample Return

The Soviets sent probes to the Moon as early as 1959[12] and in February 1966 successfully soft landed the *Luna 9* probe on the surface – the first controlled landing of a human object on another heavenly body. What was not known at the time in the West was that eleven previous attempts at a soft Moon landing had failed beginning three years earlier.[13] A taste of the rigidity of Soviet bureaucracy could be deduced from the fact that images sent back by *Luna 9* from the Moon's surface were published first in the British press – picked up there by the giant Jodrell Bank radio telescope – while the official Soviet publicity machine was still going through a laborious authorization process.

---

[10] The previous intention had been for *Apollo 8* to be another Earth orbiting mission, with the main mission of checking out the Apollo Lunar Module in space, to be followed in about March 1969 with *Apollo 9* heading to the Moon.

[11] *Apollo 8* was launched on December 21, 1968, and returned to Earth on December 27.

[12] There were four attempts to send a probe to the vicinity of the Moon in 1958, all failing because of launcher problems. There were four more launches in 1959 with three reaching the Moon – in January *Luna 1* flew past at a distance of 6,000 km, *Luna 2* impacted the Moon as planned in September, *Luna 3* took photos of the far-side of the Moon in October and the fourth experienced a launcher second stage failure (June)

[13] Four failed to reach orbit, two failed to get from Earth orbit out towards the Moon, two missed the Moon and three crashed into the Moon's surface.

There was one further soft lander, *Luna 13,* in December that year,[14] before the Soviet focus switched to a robotic probe that could return to Earth with Moon samples. As explained above, this was a fallback attempt to reduce the public impact of America being the first to send humans to the Moon's surface, if that were to happen.[15]

The program was being led by Georgiy Babakin to whom, as mentioned earlier, Korolev had passed responsibility for robotic deep space probes. His first lunar probe was the successful *Luna 9* soft lander and that gave him the support (and perhaps confidence) to take forward more ambitious concepts, such as a lunar rover (which became the eventually successful Lunokhod rover) and a sample return probe, or "scooper."

The scooper made use of the soft lander part that Babakin had developed for the lunar rover. The main challenge then was to squeeze in the required rocket motor (to return to Earth), communications, robotic arm to collect the sample and the capsule to hold the collected sample and re-enter Earth's atmosphere. To save weight there was no possibility of altering the probe's trajectory once it left the Moon's surface. The location where it landed on Earth depended on it taking off from the Moon with precisely the right speed and in precisely the right direction.

The first attempt to automatically bring Moon samples back to Earth was launched in June 1969, just a month before *Apollo 11* was due to carry Armstrong, Aldrin and Collins to the Moon. The Proton rocket had failed to reach orbit in its four previous attempts, and despite assurances by Chelomey and Glushko, the same thing happened again. The fourth stage failed to ignite, with the entire rocket and its payload ending up in the Pacific Ocean.

The Soviet Union had one last throw of the dice in the hope of bringing Moon samples back before *Apollo 11*. This time the Proton rocket worked perfectly and placed *Luna 15* on course for the Moon on July 13 (three days before the launch of *Apollo 11*). It fired its rocket to place it in orbit around the Moon on July 17 and should then have descended to the surface on July 20 after making corrections to its trajectory to ensure it landed in the correct location. The Moon's gravity field was poorly known at the time and as a result an extra 18 hours were spent trying

---

[14] The intermediate Luna missions 10, 11 and 12 were placed in orbit around the Moon to photograph its surface and measure its gravitational field and other physical parameters. *Luna 10* was another Soviet "space first" – becoming the first artificial satellite of the Moon on April 3, 1966. America's *Lunar Orbiter 1* became the second artificial moon of the Moon on August 13th and the first probe to photograph the Moon from orbit – *Luna 11* was the first Soviet probe to photograph the Moon while in orbit around it arriving there just 2 weeks after *Lunar Orbiter 1*. There were five U. S. lunar orbiter spacecraft in total photographing the surface to help locate suitable landing sites for the Apollo missions to come.

[15] The first U. S. soft landing on the Moon was *Surveyor 1* in June 1966, four months after *Luna 9*. Between then and January 1968 there were four more successful Surveyor soft landers and two failed ones.

to compensate for the unplanned changes in trajectory that the gravity field caused. This meant that the actual attempt to land *Luna 15* took place just two hours before Armstrong and Aldrin blasted off from the Moon's surface, having finished their mission. Despite the extra time spent preparing, *Luna 15*'s descent was off-course, and it ended up smashing into Mare Crisium, one of the giant craters in which there is an excess of gravity.[16]

The Soviet misinformation machine sprang into action and declared that "*Luna 15*'s research program had been completed and the spacecraft had reached the Moon in the pre-set area" ([3], p. 696).

The Soviet spin on their space program became that they had always focused first on building a space station in Earth orbit as a steppingstone to exploration of the Moon and the planets, and second to use robotic probes to obtain samples from the Moon. The lunar scooper program therefore continued as evidence of the validity of this second strand.

Three more failures followed before the first successful return of Moon dust a year later by *Luna 16* on September 24, 1970 (see Fig. 8.9). It carried to Earth about 4 ounces (101 g) of soil from the Sea of Fertility, which were eagerly analyzed by Soviet scientists. Three grams were exchanged with NASA in return for three grams each from the *Apollo 11* and *12* samples. Despite the small size of these samples they could be analyzed in depth using electron microscopy and provided useful scientific information to NASA about an area of the Moon not visited by the Apollo astronauts.

Fig. 8.9. Fifteen months after *Apollo 11*, *Luna 16* became the first robotic probe to return samples to Earth from another body. The spherical return capsule visible at the top of the 13-foot (4-m) -high stack was just 20 inches (50 cm) wide and contained the sample in a sterile sealed container, with drogue, main parachutes and a radio transmitter, protected by a heat shield, on the lower (heavier) side. (Illustrations courtesy of NASA/National Space Science Data Center.)

---

[16] Even if *Luna 15*'s mission had gone exactly to plan it would have returned to Earth two hours after *Apollo 11* splashed down.

The sample return spacecraft weighed more than 5 tons when launched from Earth and 1.8 tons when it landed a week or so later on the Moon. It spent four days reaching the Moon, then about four more in orbit around it before a stop-and-drop firing of its rocket engine to fall down to the surface – slowed at 2,000 feet (600 m) altitude by a final burn of its engine and then cushioned for the last few meters by two smaller engines to give an impact speed of 5 mph (9 kmph). Samples were obtained from 14 inches (35 cm) below the surface by a robotic arm fitted with a drill and placed in the return capsule, which was then sealed. The return capsule was lifted off the Moon's surface and directed towards Earth by its ascent rocket, rotating slowly throughout its four-day return trip so as to avoid overheating (the same "BBQ roll" used by Apollo spacecraft). Separated from its rocket motor the spherical capsule weighed about 75 pounds (34 kg) as it survived the intense heat of atmospheric re-entry and the very high g-forces (up to 350 g) as it decelerated from 24,500 mph (11 km/sec) so that it could finally descend under its parachute to a soft landing just 20 miles (30 km) from its planned destination in Kazakhstan.

There were five more lunar scooper missions of which two succeeded – *Luna 20* in February 1972 and *Luna 24* in August 1976, this last involving a sample drilled out from 7 feet (2 m) below the surface. *Luna 20*'s return journey came close to disaster in that the capsule almost landed in the Karkingir River in Kazakhstan during a snowstorm, but luckily landed instead on an island in the middle of the river.

All three Luna landing sites were near the eastern rim of the Moon as seen from Earth[17] (see Fig. 7.3), but their samples proved to be surprisingly different from each other. *Luna 16*'s samples were largely made up of basalt, which is the original lava of the Moon's surface, and were rich in iron, titanium and silicon compared to typical basalts on Earth. By contrast *Luna 20*'s sample was mainly feldspar rather than basalt, rich therefore in aluminum, silicon and calcium. Finally, *Luna 24* landed inside the giant Mare Crisium (Sea of Crises), and its samples seem to have mainly comprised material thrown up from perhaps 100 meters below the Moon's surface by the impact of an asteroid that left a 4-mile (6-km) -wide crater about 11 miles (18 km) from *Luna 24*. The resulting samples displayed a wide range of chemical forms, with perhaps the most distinctive feature being an almost complete absence of titanium. Because of this chemical diversity between the three Luna landing sites they have provided a significant addition to the much larger quantity of material brought back by the six Apollo missions ([3], p. 739; [4], pp. 181-191; [5], pp. 94-96, 107, 136).

---

[17] The right-hand edge of the Moon as seen from northern latitudes on Earth; the left-hand edge from southern latitudes; the top as seen from the equator.

The three successful Luna missions returned a total of 326 grams (11½ oz) of material from three locations on the Moon's surface. For comparison, the Apollo missions returned 1,200 times more: 382 kilograms (842 lbs or 13,472 oz) from six locations. The quantity of Moon rock and dust returned by the Luna probes seems minuscule compared to the third of a ton returned by the Apollo missions, but in terms of ounces returned per dollar spent, which was better value?

A declassified 1969 Central Intelligence Agency report [6] estimated that the Soviet robotic lunar and planetary programs (excluding launcher costs) had cost $2.3 billion up to 1969 in current money. In the period up to and including 1969, there were 70 Soviet launches of robotic probes to the Moon and beyond, of which 15 succeeded (See [7]). So as a very rough average figure, each of the 70 Soviet missions cost $33 million.

There were a total of eleven Luna sample return (lunar scooper) missions of which, as we saw above, three were successful. At a cost of $33 million per mission, that puts the cost of the lunar scooper program at about $360 million (excluding the cost of the launchers). They returned a total of 11½ ounces of Moon dust to Earth, making the *cost per ounce of Moon dust about $31 million.*

NASA reported to Congress in 1973 that the Apollo program had cost $25.4 billion (including the cost of developing the launcher). That resulted in the return of 842 pounds (which is 13,472 ounces) of Apollo Moon dust and rock to Earth, making the price of each ounce $1.9 million – which sounds like good value for money compared to the $31 million per ounce Soviet price.

The Soviet Union claimed that robotic technology was a more efficient way to perform lunar exploration than sending humans. The above calculations do not support that conclusion! Thus not only did the Soviet Union lose the race to put men on the Moon, their fallback plan to return Moon samples robotically ended up costing more per ounce of "Moon dust" than the hugely expensive Apollo program. We will return to this robotic versus human value for money debate when we look to the future in Chapter 10.

The lunar scooper program was, however, a technical tour de force given the limited technology available to Babakin's team. Hard choices were made that normally would have been considered too risky. For example, to save weight, the return to Earth of the capsule containing the precious Moon dust was entirely passive, and so its landing site on Earth was determined by the accuracy of its takeoff from the Moon plus its location on the Moon – there could be no corrections along the way. In turn that meant that the landing site on the Moon had to be within about 10 kilometers of the planned location. The descent to the Moon's surface was just as difficult as for the Apollo astronauts but without the benefit of a pilot onboard, so the need to land at a predetermined location was a major challenge.

110   **The Other Competitor in the Race**

The *Luna 17* Moon rover was another Soviet technology success on the Moon, whose impact was diminished by coming 16 months after *Apollo 11*. The *Lunokhod-1* rover, as it was called (see Fig. 8.10), landed on the Moon on November 17, 1970, and spent 10 months traversing the Sea of Rains[18] – way beyond its expected life of 3 months. The Moon's surface is one of the most difficult environments in the Solar System in which to survive, because of the two-week duration of each night. *Lunokhod*-1 used a nuclear power source[19] to keep its batteries charged and equipment warm throughout the long cold lunar night. The ¾ ton, eight-wheeled, SUV-sized *Lunokhod-1* traveled more than 6 miles (10 km) across the Moon's surface, steered by two teams of five "sedentary cosmonauts" working in two shifts from mission control in Crimea in the Ukraine[20] analyzing the chemistry of the soil, drilling into it to analyze its mechanical properties and taking panoramic photographs as it went. The sedentary cosmonauts had to cope with the four-second delay between sending a command and seeing the result on the TV picture,[21] helped by the fact that the top speed of the rover was about 100 yards (100 m) per hour.

Lunokhod-2 (*Luna 21*) followed two years later (January 1973), traveling more than 25 miles (40 km) in the lunar highlands[22] near where the *Apollo 17* astronauts had said man's goodbye to the Moon a month earlier. Stereoscopic panoramic TV imagery gave improved information about the surface compared to *Lunokhod 1*. Three months into its journey around the 35-mile (55-km) -wide crater[23] inside which it had landed it began exploring a region littered with rocks 2 to 3 yards (2 to 3 m) in size on the edge of a major canyon that stretches for about 15 miles across the crater floor, and also close to the mountainous rim of the crater (see Fig. 8.11). After a month spent in that region the rover was commanded to leave, but in doing so it rolled into a crater so that dust covered its solar panels, with a resulting damaging effect on temperature control and energy generation. Attempts to recover the rover failed and the mission ended ([4], pp. 191-201; [5], pp. 97-98, 113-114).

---

[18] Officially called Mare Imbrium, this is the largest dark area on the front side of the Moon and perceived (at least in northerly latitudes) as the left hand eye of the Man in the Moon.
[19] 11 kilograms of Polonium 210.
[20] Each team was comprised of a commander, driver, navigator, engineer and radio operator. Crimea is now under Russian control, although still claimed by Ukraine.
[21] About two and a half seconds for the radio signals to travel to and from the Moon plus delays in the control panel executing the lever movements of the "drivers."
[22] On the edge of Mare Serenitatis (Sea of Serenity), the right hand eye of the Man in the Moon for those in northerly latitudes, and about 110 miles (180 km) north of the *Apollo 17* landing site.
[23] Its official name is Le Monnier crater.

The Other Competitor in the Race 111

Fig. 8.10. Full-size engineering model of the ¾ ton, 7½ foot (230 cm) long nuclear powered *Lunokhod-1* rover that explored the Moon for ten months in 1970-71. The rover was controlled from the Soviet Union, demonstrating a form of Moon exploration that is still highly relevant today. (Illustration courtesy of Brian Harvey. Used with permission.)

In Chapter 10, when we look at the future of humans on the Moon, these impressive early Soviet robotic exploration missions (scooper and rover) will provide a useful reference.

In addition to the two successful Lunokhod missions, there had been an earlier unsuccessful one. The first attempt to send a Lunokhod to the Moon failed in February 1969 when its Proton launcher exploded a minute into its flight. That launch took place just two days before the first test flight of the giant N1 rocket (see later in this chapter) with the intention that the spacecraft carried to the Moon by the N1 would photograph the Lunokhod rover on the lunar surface ([3], pp. 679–681).

Postscript: After the *Luna 24* sample return probe in August 1976 it would be fourteen years before another robotic probe made its appearance – Japan's *Hiten* spacecraft that first made several flybys of the Moon and then went into orbit around it, and was finally deliberately crashed into the Moon in 1993.

112 The Other Competitor in the Race

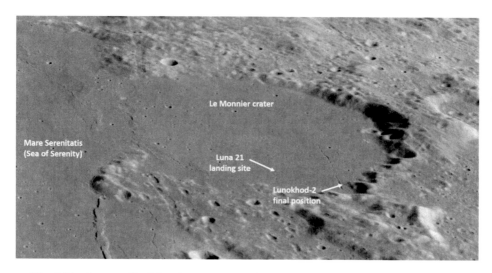

Fig. 8.11. Image of Le Monnier crater taken by *Apollo 17* in January 1972 showing the location of the *Luna 21* probe that landed there on January 15, 1973, just 32 days after *Apollo 17* astronauts Cernan and Schmitt had taken off from their landing site 110 miles (180 km) to the south. The final position of *Luna 21*'s passenger, the *Lunokhod-2* rover, is also shown close to the long tectonic fault running north across Le Monnier, and to the wall of the crater. (Illustration courtesy of NASA/author.)

The Final Lap: Landing Cosmonauts on the Moon

The development of the N1 rocket continued throughout 1966 and 1967 under the direction of Mishin, deputy of the late Korolev. Funding was never sufficient in Mishin's view. He claimed that at its peak the N1 and the L3 spacecraft it would carry to the Moon was receiving about half the funding of the Apollo program at *its* peak.[24] Requests for funding to build a ground test facility to verify the complete first stage of the N1 were repeatedly refused, even after the first launch failure. By November 1967, the first Saturn V was launched in the United States – successfully – while at the Baikonur launch site in the Soviet Union the first mockup of the N1 was being assembled. The mockup was used to check that all the pieces of the hugely complex N1 came together. The real N1 was being constructed in the massive assembly testing building close to the launch pad, and the plan was for it to be launched in March 1968.

In addition to the development of the heavy-lift N1 rocket, the Soviets had to develop a spacecraft to carry cosmonauts to the Moon, land on its surface and return to Earth – the equivalent of the Apollo Command and Service Module and Apollo Lunar Module. A totally new spacecraft was devised for the landing on the

---

[24] Apollo cost nearly $3 billion in 1966-67; N1-L3 $1.5 billion in 1967-68 (using an exchange rate of $3 = 1 rouble).

Moon's surface, called the LK (see Fig. 8.12), while a variant of the Soyuz spacecraft was used for all the other parts of the mission, called the LOK. The reduced power of the N1 compared to NASA's Saturn V meant that the LK had to weigh less than 5 ½ tons (about a third of the weight of the Apollo Lunar Module – see Fig. 8.13). LK was therefore only able to carry a single person to the surface with a maximum stay there of 48 hours. Another compromise was that unlike the tunnel that connected the Apollo Lunar Module to the Command and Service Module, the cosmonaut had to perform a spacewalk to get from the LK to the LOK mothership and back.

Fig. 8.12. Full-size model of the Soviet LK one-man lunar lander craft – accurate from an engineering point of view although lacking some equipment that stuck out near the top. Three prototype versions were successfully tested in Earth orbit in 1970-71, but the full Moon-bound version never made it into space before being canceled. (Illustration courtesy of Andrew Gray. Used with permission.)

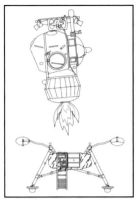 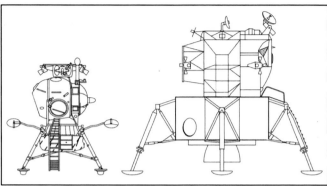

Fig. 8.13. *Left:* Artist's rendering of the LK taking off from the Moon's surface showing the separation of the main craft from the landing section. *Right:* LK and the much larger Apollo Lunar Module drawn to scale. The roughly spherical cabin of the LK in which the sole cosmonaut would stand harnessed securely in his spacesuit was about 10 feet wide by 7½ feet high (about 3 m by 2.3 m) and cluttered with piping and equipment racks. (Illustration courtesy of NASA.)

The relentless push to reduce the weight of the LK led to many design changes, and these in turn led to delays. A prototype version was built and launched into Earth orbit three times in 1970-71, each time performing well. These prototypes performed maneuvers while orbiting the Earth that mimicked a landing on the Moon, and these peculiar maneuvers were studied with interest by analysts in the West, who concluded that they were indeed preparatory flights for a human Moon landing. Full-scale units were then built to be flown on the later test launches of the N1 rocket. A lot of clever design work went into creating a compact cabin in which the lone cosmonaut could control and steer the craft despite limited scope for turning around inside. One porthole gave a view of the ground below for the final landing phase (there was enough fuel for less than a minute of hovering time) and another smaller one to the side was used for rendezvousing with the orbiting spacecraft on the return from the surface. A new spacesuit was also developed that was easy to get in and out of, and which had sufficient flexibility for both the lunar surface and for the spacewalk between vehicles.[25] 18 cosmonauts (nine teams of two) were identified for training. Spacewalk pioneer Alexey Leonov (see Fig. 8.6) would probably have been the first Soviet citizen to set foot on the Moon, with Oleg Makarov orbiting above in the LOK ([8] p176).

Things did not go smoothly with the testing of the N1 mock up and with other preparations, so it was May 1968 before the live N1 arrived on the launch pad

---

[25] LK details mainly from Siddiqi ([1], pp 488-493) and [3] 735-736. Further technical and historical details plus excellent interior photos can be found at http://www.astronautix.com/l/lk.html. Accessed 9 Oct 2018.

(see Fig. 8.14). Cracks started to appear in the giant first stage, so the whole rocket had to be returned to the assembly building for repair. Problems continued to emerge, and the delays stretched into late 1968. It was recognized by the engineers that a realistic date for the first N1 flight with humans onboard bound for the Moon was 1970 or 1971. Their feeling was that the NASA schedule would inevitably slip, thus giving the Soviets a chance still to win the race to the Moon.

Fig. 8.14. Baikonur, September 1968. U. S. Gambit KH-8 spy satellite photo of the first N1 rocket on the launch pad. The shadow of the 344-foot (105-m) -high rocket is clearly visible, making it easy for U. S. intelligence officers to calculate its size and thus its general characteristics. Problems with the N1 meant that it had to be removed from the launch pad to be modified and did not return until February 1969. This declassified version of the image is probably more blurred than the original to disguise its true quality. (Illustration courtesy of NRO.)

It was in fact February 1969 by the time the live N1 came back to the launch pad, by which time NASA's Apollo schedule was clearly on track for a 1969 Moon landing.

Soviet confidence took another beating on February 19 when an unmanned lunar rover was launched on the Proton rocket but shook itself apart after 50 seconds and exploded.[26]

Finally on February 21, 1969, the first N1 rocket was launched. It was the most powerful rocket to ever leave the ground, producing 4,590 tons of thrust. (The next most powerful rocket was the Saturn V, whose first stage produced 3,400 tons of thrust.) It took off successfully, but 70 seconds into the flight the onboard control system shut down all of the engines so that it fell to Earth about 50 miles from the launch pad. It turned out that a tiny pipe had ruptured in one of the engines, causing a fire that soon consumed that engine and others nearby, thus triggering the shutdown.

Although this first test flight was a major step forward for the N1, the general conclusion drawn by Mishin and others was that the first stage of the rocket had not been thoroughly tested, primarily because of the lack of funds to build suitable test facilities. In contrast, we saw in Chapter 2 that the first stage of the Saturn V was tested exhaustively before being assembled into the rocket, which explains why NASA's plans stayed on schedule while the Soviet plans slipped.

A few minor changes were proposed for the N1 based on the flight test, and the next live version was moved to the launch pad in late May (see Fig. 8.15). The sequence of events was eerily similar to the previous launch in that first an unmanned lunar probe was launched on a Proton rocket on June 14 but ended up in the Pacific Ocean due to a failure in the fourth stage.

With the world's media and celebrities assembling at Cape Kennedy in anticipation of the July 17 launch of *Apollo 11,* a similar but totally secret assemblage was taking place of Soviet space dignitaries at Baikonur in anticipation of the July 3 N1 launch – which, unlike *Apollo 11,* was unmanned!

The N1 lifted off at 11:18 p. m. Moscow time on July 3, 1969, rose 200 meters into the air, then fell back and exploded on the launch pad. The force of the explosion was extraordinary. A half-ton fuel tank landed on the roof of a test building 4 ½ miles (7 km) away; large windows were shattered 25 miles (40 km) away; pieces of the rocket were found 7 miles (11 km) away. Safety precautions, however, had been thorough, and there were no fatalities or injuries. The explosion was estimated to have been the equivalent of 250 tons of TNT – not quite a nuclear explosion, but certainly the worst rocketry explosion in history. Personal accounts of the events speak of the Earth and the air shaking, and of a few seconds of silence until the full roar of the launch and the explosion reached them. The following morning the clean-up team found the steppe around the launch site covered with dead animals, including birds ([3], pp. 690-692).

Soviet secrecy about their manned Moon landing plans continued. Public statements emphasized the strategy of laboratories in Earth orbit in which cosmonauts could undertake research, and robotic return of samples from the Moon. Human

---

[26] The nuclear fuel source (Polonium-210) inside the lunar rover wasn't found among the debris. The rumor is that soldiers "rescued" it and used it to heat their barracks throughout the winter.

Fig. 8.15. Baikonur Cosmodrome, early July 1969, two weeks before the launch of *Apollo 11* half a world away. On the right is a fully functional N1 rocket with a payload set to go into orbit around the Moon. On the left is a mock-up of the N1 used for rehearsing launch operations. (Illustration courtesy of Asif Siddiqi. Used with permission.)

exploration of the Moon was given short shrift, even to the point of not showing the *Apollo 11* videos to the Soviet public. One journalist recalled that the announcement of Armstrong and Aldrin's achievement was shown on Soviet TV during a break in a volleyball match between two local teams ([3], p. 697).

But the N1 program continued for another 5 years! One reason for continuing after the Americans had won the race to the Moon was to do with bureaucratic inertia in the Soviet system. (Everyone is reluctant to make a concrete decision to change an approved plan.) Another reason is that a decision to cancel such a large program as the N1 risks becoming public and thus revealing that there had after all been a Soviet program to land men on the Moon. Better to let the program meander on until a natural moment could be found not to renew it. By about 1972 another reason was that the Americans had given up on the Moon and seemed unclear about their next objective in human spaceflight; it was argued therefore that a clear and successful Soviet exploration of the Moon could restore Soviet prestige.

It took a year for the investigation into the July 1969 disaster to submit a report, with follow up reports continuing for nearly another year. The possible reasons included a failed oxygen monitor and a blockage in one of the oxygen pumps, plus an unknown reason why the control system shut down all the engines. Although not directly implicated in the failure, there were doubts raised by engineers about the adequacy of the fuel tanks and pipes.

The third N1 launch was set for June 1971 aimed at verifying the N1 launcher. Unlike the two earlier launches it did not carry a working payload. Minor technical problems kept pushing the launch back day by day. Despite a number of unresolved problems the rocket was launched on June 27, 1971, but after 51 seconds the engines shut down, and the rocket drifted on for 15 miles until it hit the ground, creating a deep crater. Another long analysis of the failed flight followed, but there was surprisingly little pressure to terminate the N1 development, perhaps because at this point many of the parts for two more N1 rockets had been procured, so the extra cost of continuing was relatively small.

The fourth (and final) N1 launch took place 17 months later, on November 23, 1972. This time it looked like it was going to be a success until just before the first stage finished its operation and handed over to the second stage. One of the 30 first-stage engines exploded, destroying the whole rocket.

A year and a half later (May 1974) the repercussions of this long list of hugely expensive failed launches finally emerged. Mishin was dismissed as head of the organization that Korolev had created, and Korolev's bitter rival Glushko was appointed in his place, also keeping responsibility for his own organization. One of Glushko's first actions was to cancel all work on the N1 rocket. He went further and ordered the seven remaining N1 rockets, two of which were ready to be launched, to be destroyed, along with all N1 technical documentation!

Summarizing the Soviet/U. S. Differences

There are several ironies in what followed after the cancelation of the N1 program, including the fact that throughout the 1970s and 1980s the Soviet Union continued the space race with America, vying to create laboratories in Earth orbit that could support many cosmonauts at a time. By comparison the United States no longer saw itself in competition with the Soviet Union (having won the Moon race) but failed to settle on an alternative space objective. Instead the United States chose to pursue technology in the form of the space shuttle. This is a reverse of roles from the 1960s, when the United States clearly specified the objective of landing men on the Moon before the Soviet Union, while the Soviets failed to provide the funding required for any of the several long-term objectives it identified and instead made a series of funding decisions for short-term political reasons.

Glushko banished the engines used in the N1 to the scrap heap, but in fact about 90 of the engines were hoarded by Kuznetsov's team. After the end of the Cold War, U. S. space companies became interested in these engines and bought most of them. Furthermore in the Soviet Union itself an improved version of the engines is now used in the Soyuz rocket.

Ironies aside, the Soviet Union benefitted in the 1950s from the interest of the military in long-range rocketry as a way to achieve strategic military parity with, or even dominance over, the United States. The technology of the first ICBM was then adapted by Korolev to send first objects and then humans into space, gaining great prestige for the Soviet Union and its political ideology. The central role of the military became counterproductive to human spaceflight in the 1960s, since there were no strong military reasons for sending humans into space.

The story of the Soviet lunar programs is one of constant lack of adequate funding for the immensely difficult technical challenges, complicated by the internal competition among the space organizations for their own pet projects, sponsored by their various favored politicians, with frequent changes of priority and policy U-turns.

The chaos of the Soviet system was evident in the 1967 decision to go ahead with three Soviet lunar programs (manned landing, manned around the Moon, and robotic sample return), each to be launched by a different (and new) rocket – a crisis-mode decision triggered by the Soviet three-year misinterpretation of President Kennedy's 1961 "Moon this decade" speech as a publicity stunt. Even worse, there continued to be attempts (each causing a delay in the program) to change the engine used by the N1 heavy-lift rocket. The decision to go ahead on three fronts was triggered by the successful launch of the American Saturn V, which demonstrated that the United States was absolutely serious about placing humans on the Moon. But instead of focusing funding and resources, the internal competition in the Soviet system led to the scattershot approach of trying all available ideas simultaneously.

The U. S. space story also begins with the military, but for a different reason than the Soviets. Imaging from space was recognized in the 1950s by the top U. S. leadership as the only reliable way to get information on military developments inside the secretive and enormous Soviet Union. The CORONA spy satellite program and the associated launcher rockets were given the go ahead by President "Ike" Eisenhower in 1955. When the Soviets sent first *Sputnik* (1957) and then Gagarin (1961) into space, the United States was able to respond quickly because of its already ongoing space developments. The recognition by President Kennedy of the public desire for American astronauts to take the lead over the Soviet Union led directly to the decision to send men to the Moon.

In stark contrast to the chaotic Soviet decision-making process the United States had NASA, created by Eisenhower to manage civil space programs. Ike was generally against such programs, but if they were going to happen, he wanted them to be managed efficiently. He saw that the first fundamental step was to designate a single government body separate from the military to be responsible for delivering such programs, and since a suitable one didn't exist, he set up what ultimately became NASA. Decisions were open and accountable so that politicians and taxpayers could see if their money was being well spent, and influence it accordingly.

President Kennedy was able to use NASA to deliver his Moon-landing commitment, separate from the military. NASA then helped the Kennedy and later administrations lobby Congress for the required funds. Progress in the United States from the one-man Mercury capsule to the two-man, steerable, dockable Gemini spacecraft to the three-man Apollo capsule and its Lunar Module lander was all part of a single strategy. The Saturn V rocket, whose full scale development began in 1964, was part and parcel of the same strategy. There was no attempt to address military requirements or to favor certain industrial groups.

The lack of an equivalent to NASA and its oversight by Congress was the Achilles heel in the Soviet system. Decisions were made for reasons of personal favoritism and short-term political gain, all without external oversight. There was no organization (other than the military) with the mandate to make long-term investment decisions in support of a national space goal. The Soviets had no strategy – other than to develop long-range missiles for their nuclear bombs. There was no civilian Soviet space program as such. Each space funding decision was made on the grounds of priorities at that moment in time, and those priorities could be changed by personal influence of top officials. Decisions were secret, so there was no public accountability.

References

1. Siddiqi, A., *Sputnik and the Soviet Space Challenge*, University Press of Florida, 2003.
2. Norris, P., *Spies in the Sky*, Springer Praxis (Chichester, UK), 2007.
3. Siddiqi, A., *The Soviet Space Race with Apollo*, University Press of Florida, 2003.
4. Harvey, B., Zakutnyaya, O., *Russian Space Probes*, Springer Praxis (Chichester, UK), 2011.
5. Siddiqi, A., *Beyond Earth: A Chronicle of Deep Space Exploration 1958-2016*, NASA SP-2018-4041, Sep. 2018.
6. Available at https://fas.org/irp/cia/product/sovmm69.pdf accessed October 9, 2018.
7. http://www.russianspaceweb.com/spacecraft.html accessed October 9, 2018.
8. Lardier, C., L'Astronautique Soviétique, Armand Colin (Paris, France), 1992 (in French).

9

Why Is It Taking So Long to Return to the Moon?

The Last Men on the Moon?

When Harrison Schmidt and Gene Cernan returned to Earth at the end of the *Apollo 17* mission they had hopes of returning to the Moon within a few years. Cernan climbed up the ladder into the Lunar Module just before midnight (Houston time) on December 13, 1972 declaring "I take man's last step from the surface, back home for some time to come – but we believe not too long into the future". He didn't realize how much history he was making (see Fig. 9.1). He would be the last human to walk on the Moon's surface for at least 50 years.

The next step in NASA's exploration of space was said to include a permanent base on the Moon, much like the scientific stations in the Antarctic.

There were some doubts about whether to continue using the Saturn V rocket or to build a new one, and many other technical issues. In addition, the Soviet Union was still expected to send humans to the Moon, so it seems bizarre that in the five decades since the Apollo missions, no humans have been to the Moon – not even to orbit around it, let alone to land on its surface. Why?

With the benefit of hindsight we can identify three broad reasons why humans have not gone back to the Moon: improvements in robotic technology, lack of political will and lack of military interest. Each of these three reasons will be discussed in this chapter, and after that we will look at the chances of humans returning to the Moon anytime soon.

122   Why Is It Taking So Long to Return to the Moon?

Fig. 9.1. The plaque left on the Moon's surface by astronauts Cernan and Schmidt in 1972 that celebrates completion of "the first exploration of the Moon." (Illustration courtesy of NASA.)

Is It Just a Question of Money?

However, let's begin by noting that the fundamental reason is one of money. It has been possible for government to justify sending humans to orbit Earth – to the International Space Station, for example – but no further. You can get to the International Space Station on a rocket that is also used to launch communications and navigation satellites, so the cost of developing the rocket has been spread over dozens, even hundreds, of launches. A rocket large enough to get humans to the Moon's surface and back has no other use, so you have to justify developing this rocket solely on the basis of human missions to the Moon.

The American government justified the high cost of developing the Saturn V rocket in the 1960s by making a manned Moon mission a strategic national goal. Congress and the American public broadly supported the decision to use their taxes to pay for that activity, because "beating the Soviets" was generally a popular idea.

In the Soviet Union, the funding to develop the N1 rocket needed for the manned Moon mission was more difficult to obtain. The politicians agreed that "beating the Americans" was important (the public had no say in this matter in the Soviet system) but then said that much of the funding had to come from the military budget. The military could see little use for the massive rocket and failed to provide adequate funds.

The logic remains the same today. The satellites we need for weather forecasting, TV broadcasting, in-car satellite navigation or air traffic control all weigh less than 10 tons[1] – many of them much less than that. However let's not oversimplify; satellites such as those for TV broadcasting are often located 22,000 miles (35,000 km) high in order to appear stationary from Earth. They are in a sweet spot where an object is moving at the same speed as Earth's rotation and staying fixed above a point on the surface – the so-called geostationary orbit. Getting a 1-ton satellite to that altitude requires a rocket about twice as powerful as required to put the satellite at 400 miles altitude. In practice rockets are needed to launch geostationary satellites weighing up to 10 tons or so, and satellites weighing up to about 20 tons into a 400-mile-high orbit, and these two tasks are roughly equivalent. For comparison, recall that the Saturn V that took Apollo to the Moon could place about 130 tons into an orbit 400 miles high.

Table 9.1. Currently operational heavy-lift space launchers. (May 19, 2019, alphabetical order).

Launcher	Country of Manufacture	Mass to Low Earth Orbit (tons)	No. of Successful Launches
Angara A5	Russia	24.0	1
Ariane 5 ECA	Europe	21.0	70
Atlas V 551	United States	18.5	9
Delta IV Heavy	United States	28.8	11
Falcon 9	United States	22.8	49
Falcon Heavy	United States	63.8	2
GSLV Mk III	India	10.0	2
H-IIB	Japan	16.5	7
Long March 3B/E	China	11.5	44
Long March 5	China	25.0	2
Long March 7	China	13.5	2
Proton M/M+	China	23.0	104
Soyuz-2.1A/B	Russia	8.2 (9.0)	60
Soyuz –FG	Russia	6.9	68

[1] There are at least three different weights that are pronounced "ton." For brevity, I use the word "ton" to signify a weight of 1,000 kilograms (about 2,205 lbs) instead of "tonne" or "metric ton." Note that in the United States and Canada, "ton" usually means 2,000 pounds, while in the rest of the world it usually means 2,240 lbs.

Table 9.2. Heavy-lift space launchers now retired.

Launcher	Country of Manufacture	Mass to Low Earth Orbit (tons)	No. of Launches
Saturn V	United States	130.0	13
Space Shuttle	United States	24.4	135

Table 9.1, from Wikipedia [1], lists the rockets currently in use to place large objects into space. One, the Falcon Heavy, can place more than 50 tons into a low Earth orbit (200 to 400 miles high). As of May 2019 it has been launched just twice, with a third launch imminent. We will return to this rocket in Chapter 10 when discussing the future.

All the other rockets in Table 9.1 top out below 30 tons; six have a capacity between 20 and 30 tons. The only launcher that carries astronauts and cosmonauts to the International Space Station is the Soyuz-FG, whose capacity is not quite 7 tons. For completeness, Table 9.2 shows the capability of two launchers that are no longer available: the space shuttle, with a capacity of 24 tons, and the Saturn V.

The wonders of microelectronics are driving down the size of many satellites, and in the opposite direction there are also tendencies to make them heavier in order to achieve economies of scale. One example among hundreds of the lightweight trend is the *Carbonite-2* satellite launched January 20, 2018, that takes short videos from space and weighs just 100 kilograms[2] In the 1990s such satellites were only affordable by the military and weighed several tons. An example of the opposite trend is the *Telstar-19V* satellite launched on July 22, 2018, into geostationary orbit. This is the heaviest commercial communications satellite to-date, weighing just over 7 tons at launch.

The military, too, can get by with satellites weighing not much more than 10 tons, and usually much less – spy satellites, eavesdropping satellites, stealthy communications satellites and more. Details of military missions are by their very nature difficult to obtain or to verify, but in recent years the only American satellites requiring the most powerful of the regularly used rockets in Table 9.1, the Delta IV Heavy, were (probably) a couple of U. S. spy satellites.

There are a few exceptions to the above generalizations, which we will return to in Chapter 10, where the subject of rockets that can launch heavy weights into space is an important topic in looking to the future. But right now, the 100 or so launches every year of commercial, human, scientific and military spacecraft get by with rockets that can lift less than 30 tons into orbit. This is a long way from a rocket that can place 130 or so tons in orbit as needed to get the Apollo missions to the Moon and back.

---

[2] Examples of the videos can be found on YouTube at https://www.youtube.com/watch?v=VOZE6lvpzYo.

Moore's Law

Perhaps the most obvious reason why human flights to the Moon ended in the 1970s is that since then we humans have been outperformed by machines. We continue to breath oxygen, drink water and eat food in the same quantities as we did then. We still sleep about 8 hours a day, we weigh the same (or more) and take up the same space (or more). On the other hand electronic machines have gotten smaller and cheaper while performing the same amount of work. They use much less electricity than they did before and take up a lot less space.

We see the evidence of this ever-increasing power of electronics in our cell phones, flat-screen TVs, digital cameras, video game consoles and the like – they get better and better despite getting smaller and lighter, and tend to cost much the same (perhaps a bit more in the case of a smart phone, but less in the case of a laptop). The rooms full of computers and storage devices at Apollo Mission Control in Houston in 1969 are outperformed by the computer in a modern smartphone. And the smartphone doesn't need the special electricity generation station to provide the tens of kilowatts of power needed not only by the 1969 computers but also by the industrial-scale air conditioning that prevented them from bursting into flames.

The boss of computer chip manufacturer Intel, Gordon E. Moore, spoke about this trend in the 1960s – a doubling of the power of computer chips every two years or so. As the trend continued into the 1980s and beyond it became known as Moore's law in honor of its first annunciator (see Fig. 9.2). It's still going strong today, so if you do the math you find that 50 years after *Apollo 11* (1969) an electronic device has typically increased its power about 30 million times – and counting.

It was difficult to persuade governments to fund the cost of placing humans in space back in the 1960s, when robotic spacecraft could perform pretty well at a small fraction of the price. The argument is now 30 million times harder to make – and getting harder as each year goes by. The Lunokhod rover and Luna sample return (lunar scooper) probes that the Soviet Union did eventually successfully get to the Moon (see Chapter 8) would be much more capable and lighter today thanks to our electronics technology.

You might wonder if the rockets that get you into space have come down in price like electronics. In fact, today's most successful rockets use pretty much the same technology as in the 1960s, in some cases of the 1940s. They have improved by perhaps one or two percent per year over the 50 years, unlike electronics that improve that much every couple of weeks!

There's No One to Compete With

A second reason no one has been back to Moon since *Apollo 17* astronaut Gene Cernan in December 1972 is that we are no longer in a high stakes competition to do so.

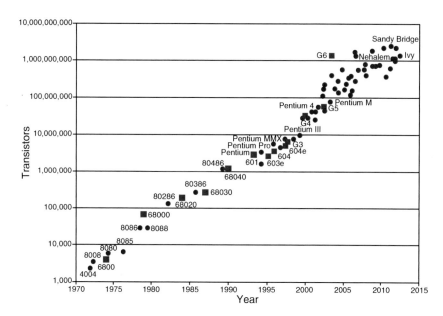

Fig. 9.2. The number of transistors in a computer Central Processing Unit (CPU) chip doubles every two years in line with Moore's law – note the log scale on the vertical axis. The law applies not only to the speed of computers but also to the size of computer memory and the number of pixels in a digital camera.

As explained in earlier chapters the decision to fund the enormous expense of "sending men to the Moon and bringing them back safely" was taken to demonstrate to the American people and the world that the United States was the number one superpower. Going to the Moon was seen as demonstrating that the Soviet Union's apparent leadership in space was a thing of the past. The Soviets had launched the world's first manmade satellite, *Sputnik-1,* in 1957 and followed with the first human in space, Yuri Gagarin, in 1961. Sending a man to the Moon was seen as a clear way of demonstrating that the United States led in space – as in pretty much every other form of technology.

The United States and the Soviet Union were in a state of near conflict in several parts of the globe – from Cuba on the U. S.'s own doorstep to Korea in east Asia, and from Berlin in Europe to Taiwan off the coast of China. Both superpowers felt the need to demonstrate to the world the merits of their form of government – democracy in the United States; communism in the Soviet Union. This competition between ideologies was not limited to the military sphere but often extended to every aspect of life: sport (i. e., number of gold medals at the Olympic Games), education, health, agriculture, economics, housing, transport and more. Technology was where the United States felt it had the most obvious lead, but this had been cast into doubt by the Soviet Union's early space triumphs. Landing on the Moon would re-establish America's technological leadership beyond dispute (see Fig. 9.3).

Why Is It Taking So Long to Return to the Moon?

Throughout the 1960s the United States and the Soviets extended their space activities – launching two astronauts at a time, performing space walks, rendezvousing in space with another spacecraft and so on. In the United States, the president, the Congress, the media and the public gave strong support to the idea of overtaking the Soviets in space. NASA's funding became one of the largest items in the federal budget without any serious dissent.[3]

Fig. 9.3. September 24, 1962, Rice University, Houston, TX. President Kennedy defused criticism of his expensive commitment to the Apollo program with the memorable line: "We choose to go to the Moon in this decade and do the other things, not because they are easy, but because they are hard." (Illustration courtesy of NASA.)

And of course the United States *did* win the race. The money spent on Apollo achieved what it had set out to do. America was the space superpower. The Soviet Union was very competent in space, but not to the extent of sending humans to the Moon. In the mid-1970s the two superpowers even agreed to perform joint space missions, most noticeably the Apollo-Soyuz mission in which U. S. and Soviet astronauts rendezvoused in space and performed joint experiments.

Once the race was over, the president, and many in Congress, the media and the public *did* question the sums of money sought by NASA. Fig. 5.1 showed the

---

[3] At its peak (1965-66) 4% of the federal budget was required to finance the Apollo program - see Chapter 5. There seems little political appetite for a commitment on that scale for a mission to return to the Moon.

precipitous drop in NASA's budget after Apollo. Further human flights to the Moon were quickly off the agenda, although vaguely foreseen at some point in the future. The same was true for human flights to Mars – the "obvious" next destination according to the space enthusiasts.

Twenty years later a committee of inquiry into America's objectives in space (the Augustine committee, chaired by industrialist Norm Augustine) posed the question "Would we be content with a space program that involved no human flight?" The report said that "Our answer is a resounding "no" [2], but also said that there was a "lack of a national consensus as to what should be the goals of the civil space program." It concluded that NASA's human spaceflight program "should be tailored to *respond to the availability of funding*"[4] even if its long-term goal was to send humans to Mars – "adhering to a rigid schedule" was to be avoided. Contrast that with the strategy that worked for Apollo: "Get to Moon by 1969, whatever the cost."

The 1990 Augustine report and many other reports before and since have argued for international collaboration as the way to perform human space exploration in the future. "This is a challenge that could be constructively shared among a number of nations" was how Augustine phrased it.

Augustine's advice to collaborate flies in the face of history. Competition has underpinned space programs since the very beginning. In the Soviet Union in 1955 the Soviet Politburo approved the development of Sputnik following reports in the media that the United States was giving priority to its satellite program [3]. And competition between the United States and the Soviets fueled the Moon race.

Competition between the two superpowers during the Cold War was especially intense in the military sphere. After the Second World War, the United States was the only country possessing the atomic bomb – generally referred to as a nuclear weapon. The Soviet dictator, Stalin, was determined not to allow the United States to remain the only country with such weapons. He gave priority to developing "the bomb," and in 1949, much sooner than expected by the United States, the Soviet Union exploded an atomic bomb helped by information supplied by several Soviet spies in the U. S. atomic bomb program. The United States responded by giving priority to developing the even more powerful hydrogen bomb, which it did by 1952. The Soviets had anticipated this U. S. reaction and had begun their own hydrogen bomb effort, exploding the first one in 1953, less than a year after the United States.

This competitive escalation of the weaponry of the two superpowers continued and spread especially into those weapons that could deliver a nuclear bomb to the enemy: aircraft, submarines and above all, missiles. The United States had certain advantages over the Soviet Union in that it had allies all around its rival in Europe,

---

[4] Emphasis added.

the Middle East, and southern and eastern Asia, from which short-range missiles could reach deep into Soviet territory. The United States also had a large fleet of long-range bomber aircraft that could reach the Soviet Union. The Soviets lacked allies close to the U. S. heartland and also lacked long-range bombers, so missiles became a top priority for them – long-range missiles with globe-spanning range (so-called ICBMs) or shorter range missiles that could be launched from submarines under the ocean off the U. S. east and west coasts (so-called Submarine Launched Ballistic Missiles, SLBMs).

Building on their own indigenous rocket expertise (the Soviets produced 10,000 of the Katushya rocket artillery launchers starting in 1941 that were much feared by the Germans) and on technology and experience imported from Germany after the war, a Soviet long-range missile was soon under development. As recounted in Chapter 8, Korolev retained and improved the kerosene and liquid oxygen fuel combination used by Germany for its V-2 rocket and by 1957 had built and successfully tested the R-7 missile with a range of over 5,000 miles. He then modified the missile to carry Sputnik into space that same year, initiating the Space Age.

Some of the political reactions to Sputnik in the United States were almost hysterical, and these had the effect of triggering further developments in the Soviet Union. For example, the leader of the U. S. Senate, Lyndon B Johnson (later vice-president, and later still president) claimed that "control of space means control of the world." He used his influential position in the Senate[5] to argue for the use of space in national defense, and supported big increases in NASA's budget. The head of research in the Department of Defense, Herbert York, also argued publicly for a strong military space program.

These two influential American voices helped persuade the Soviet leadership that its space program needed to be more military and not just scientific and diplomatic. A direct consequence of this was the creation of a secret "space plan" in 1960 that included the development of a rocket capable of placing 100-ton payloads into orbit with largely military objectives ([4], pp. 237–240). Although funding was not made available to actually deliver the ideas in the plan, it provided the background justification for what became the N1 Moon program launcher for the next decade.

Competition with the United States influenced the urgency for the Soviets to launch the first man into space, Yuri Gagarin. Korolev was not only developing the Vostok capsule to carry humans into space, he was building interplanetary probes and, crucially, military surveillance (spy) satellites. The Zenit satellites, as they were called, rapidly increased in importance to the Soviet military as they saw the United States testing its CORONA spy satellites. In January 1960 Korolev's boss,

---

[5] Chairman of the Senate Aeronautical & Space Sciences Committee as well as Senate Majority Leader.

Dmitriy Ustinov,[6] warned him that there was no goal more important at the present time than the Zenit reconnaissance satellite program. However, barely six months later the progress in the American Mercury manned space program caused a reversal in Soviet priorities.

Having kick-started the space race with Sputnik in 1957 and having consistently held the lead with one space spectacular after another,[7] the Soviet Union was forced to try and maintain its preeminence. The United States published the schedule for the first American astronauts in space, which allowed the Soviets to plan to do so first. Thus in the autumn of 1960 Soviet Premier Nikita Khrushchev and his deputy Frol Kozlov threw the weight of the whole Soviet apparatus behind the Vostok manned program ([4], p. 255), ordering Korolev to launch a (Soviet) man in space by December that year. There were some delays to this very tight schedule, not least the horrendous explosion on October 24, 1960, of a long-range missile (the R-16 developed by Korolev's rival, Yangel) on the launch pad at Baikonur that killed about 130 people, including Marshal Nedelin, Commander in Chief of the Strategic Missile Forces, which delayed all further activity at Baikonur for several weeks. Yangel himself narrowly escaped death, having gone to a bunker for a cigarette when the explosion occurred ([4], pp. 256–258).

But eventually, on April 12, 1961, Gagarin was launched in *Vostok-1*, entering the history books and crowning three and a half years of Soviet space dominance. He just beat Alan Shepard's suborbital (up and down) flight into space on May 5.

Competition is not always a positive factor. We saw in Chapter 8 that competition between industry groups in the Soviet Union resulted in a fragmented program with many changes of priority. The Soviets lacked the market processes to turn competition into a positive influence. The approval of multiple programs aimed at similar objectives prevented scarce funding and skills being focused on a single objective, with the result that the high-level objective of beating the United States to the Moon was not realized. As space historian Asif Siddiqi put it, "In the centralized and socialist Soviet system, with resources restricted by the needs of the defense sector, [competition] gave rise to chaos" ([4], p. 242). Whereas in the United States, competition tends to generate innovation, as industry groups vie to outperform their competitors and capture increased market share.

The track record of collaboration is less glorious. Collaboration in human spaceflight since the turn of the century has primarily involved the International Space Station (ISS). At a cost to date of more than $100 billion, the ISS has certainly followed the suggestion in the Augustine report to use the available funding, and lots of it. The achievements of the ISS are less easy to identify than were those of the Apollo missions, which in today's money also cost a little over $100 billion [5].

---

[6] At that time, Chairman of the Military-Industrial Commission.
[7] First animal in space (Nov. 1957), first animal recovered from space (Aug. 1960), first probe to hit the Moon (Sept. 1959), etc.

Back in the 1970s NASA was said to be 20 to 30 years away from sending humans to Mars. After spending the equivalent of the Apollo budget on the ISS, NASA is *still* 20 or more years away from that goal.

Competition can get things done in other scientific and technical disciplines. Look at the human genome project, where competition from a private sector organization led by the entrepreneurial Craig Venter spurred the generously funded public sector collaborative research group into finishing the project at least two years earlier than would otherwise have been the case. And in an earlier episode of the same story, back in the 1950s, the discovery of the structure of DNA by Crick and Watson at Cambridge University was given urgency by their perception of being in a race with Linus Pauling of the University of California.

Competition doesn't just spur on public sector or university groups. In the 1960s IBM was the world's most successful computer company. Its legendary Chief Executive Thomas J. Watson, Jr., was asked by a shareholder at the company's annual general meeting how he could justify funding the same research at two different IBM research laboratories. He replied that given the importance of the research objectives, "How could you justify *not* funding it at two laboratories?" Watson understood that a certain amount of competition sharpens people's minds so that they focus more clearly on achieving their objectives. In IBM's case they were so far ahead of the competing computer companies at the time that they had to supply their own competition internally.

Even in the United States, competition without some regulating force such as the market can be very destructive. The root cause of the Cold War nuclear arms escalation was probably the competition *within* the United States between the three armed services, accentuated by the American political system. The result was that the United States and the Soviets each ended up with enough nuclear weapons to destroy all life on the planet several times over.[8]

The competition that fueled the 1960s Moon race was political – each of the world's two superpowers seeking to demonstrate the superiority of its way of life. Such competition still exists, although perhaps not in quite such a black and white form. Take for instance the competition between the United States and China today. The underpinnings of this competition as seen from the United States are commercial:

- China has taken over the U. S. mantle as the world's leading manufacturer.
- There is a huge excess of U. S. imports from China over its exports to China. In 2017, the $40 billion positive balance in services was swamped by the $376 billion negative balance in goods.[9]

---

[8] See [6], pp. 124–125 for a discussion of this point.
[9] See Exhibits 20a and 20b of [7].

From the Chinese perspective, the United States is a rival not only in commercial terms but also politically:

- China's construction of artificial islands in the South China Sea is constantly challenged by U. S. Navy ships seeking to preserve (U. S. point of view) or establish (China's point of view) freedom of movement on the high seas.
- The U. S. commitment to defend Taiwan from Chinese attack is seen by China as U. S. interference in an internal Chinese matter.

Competition in space between the United States and China has been largely indirect. Both countries have a significant number of military spacecraft in orbit monitoring military activities of interest to them. China felt the need in 2007 to demonstrate that it could destroy satellites in space if it wanted to – it fired a missile at one of its own broken down weather satellites, smashing it into thousands of pieces, later admitting that this was indeed a test of an anti-satellite weapon as had been reported in the Western media.

A Chinese spokesman went on the say that "This test was not directed at any country and does not constitute a threat to any country." Well, actually, it *does* present a threat to *every* country with spacecraft in orbit, and that includes the number one space power – the United States.

China has continued with its aggressive space tactics. A satellite labeled *Shi Jian-17 (SJ-17)* was launched into geostationary orbit 22,000 miles (36,000 km) high in November 2016, located over the Marshall Islands in the central Pacific. It was described as an experimental communications technology satellite (the label Shi Jian means experimental). Commentators in the West focused their attention on the rocket that carried *SJ-17* into orbit, which was the first launch of the new generation Long March 5 rockets (see Chapter 11). The significance of its payload, *SJ-17,* was largely overlooked, at least initially.

During the first six months, *SJ-17* approached very close to another Chinese satellite, *ChinaSat-5A,* which had been launched in 1998 and thus close to the end of its useful life. *SJ-17* got within 4 km of *ChinaSat-5A* and then settled into a nearby orbit at a distance of about 100 kilometers. With the benefit of hindsight we can see that this exercise was one of testing the ability to approach and then remain close to another satellite in geostationary orbit.

In April 2017, six months after its launch, *SJ-17* started wandering along the geostationary arc – still 22,000 miles high and thus in a 24-hour orbit, but moving across the sky and thus not quite "stationary" as seen from the ground. The drift across the sky has been halted half a dozen times for periods ranging from 7 days to 3 months, and the drift has ranged from the central Pacific (178° W longitude) to east Africa (40° E longitude) [8]. In the past, Russia has been accused of placing one of its Luch geostationary satellites close to commercial satellites in order to listen in to communications on the latter, presumably for national intelligence reasons rather than commercial ones. It is possible that this is the purpose of China's *SJ-17*, but many of the satellites it has got fairly close to have been Russian in origin, which at the very least suggests a lack of trust between the two countries.

The United States, too, has been active in "inspecting" other countries' geostationary satellites, and has made public statements on the matter. In 2014 head of U. S. Air Force Space Command Gen. William Shelton announced that the United States will monitor objects in geostationary orbit using a system called Geosynchronous Space Situational Awareness Program (GSSAP) (see Fig. 9.4). The first pair of GSSAP satellites were launched on July 28, 2014, and a second pair two years later. The U. S. Air Force website states openly that their purpose is to get up close to other people's satellites: "GSSAP satellites operate near the geosynchronous belt and have the capability to perform Rendezvous and Proximity Operations (RPO). RPO allows for the space vehicle to maneuver near a resident space object of interest, enabling characterization for anomaly resolution and enhanced surveillance." The techniques employed by GSSAP were trialed by the Mitex pair of experimental satellites launched in 2006 [9, 10].

Fig. 9.4. Artist's rendering of the twin satellites in the U. S. Geosynchronous Space Situational Awareness Program (GSSAP) that inspects and monitors satellites in geostationary orbit. Russia and China have satellites that appear to do something similar – getting close to another geostationary satellite, perhaps to listen to its communications traffic. So far, this activity has not resulted in damaging another spacecraft, but it seems only a matter of time before that happens. (Illustration courtesy of the U. S. Air Force.)

President Trump's threat of a U. S. trade war with China highlights the potential for competition between the two countries to intensify. As we saw during the Cold War, competition in space can become a non-violent arena for competing superpowers to demonstrate their prowess. Will the China-U. S. rivalry follow that example?

Competition between countries in space can also take on a very commercial flavor. China has provided commercial satellites to a few countries such as Nigeria as part of a larger trade deal with that country. Seen from the West this looks like China subsidizing the spacecraft price in return for the trade deal. China of course talks about establishing a new relationship with the recipient country that is bigger than any one sale. Strengthening diplomatic relations can also be achieved by inviting a citizen of another country to visit your space station – a device used several times by the Soviet Union during the Cold War. An invitation to have one of your citizens participate in a Moon expedition could be a powerful diplomatic tool if and when China (or any other country) gets close to having a viable Moon program.

Let's not overlook good old-fashioned commercial competition. A competition for unmanned lunar rovers was set up by Google in 2007: the Google Lunar X-prize. About 20 teams around the world tried to "land a robotic spacecraft on the Moon, travel 500 meters, and transmit back to Earth high-definition video and images" (See [11]). Five teams won intermediate prizes worth $5.25 million, but eventually the clock ran down, and the $20 million first prize went unclaimed.

Some of the Google Lunar X prize competitors are still active, one example being Moon Express, which has received some NASA funding to send robotic probes to the Moon. It argues that "The Moon is Earth's 8th continent, with precious resources that can bring enormous benefits to life on Earth," which underpins its long-term objective of "returning to the Moon and unlocking its mysteries and resources for the benefit of humanity."

A smaller scale European alternative to the Google Lunar X prize emerged in September 2018, spearheaded by aerospace giant Airbus. Monetary prizes will be won by teams that identify and undertake technology development for sustainable lunar exploration in an international competition called The Moon Race. The "sustainable" label refers to techniques such as manufacturing an object or structure using lunar materials, generating energy to light the lunar night, producing a bottle of lunar water, or building and operating a lunar greenhouse [12].

A more conventional form of competition is possible given NASA's commitment under the Trump Administration to increase the role of industry in manned exploration of the Moon (and Mars).

Jim Bridenstine had barely taken up his role as head of NASA when this policy was put into concrete action. NASA's planned lunar rover mission, Resource Prospector, was stopped, and will re-emerge only when winners have been chosen from a competition to provide "Commercial Lunar Payload Services."

Something similar happened in the George W. Bush administration when commercial suppliers were brought in to deliver cargo to the International Space

Station. Two suppliers[10] were chosen to maintain an element of competition (as well as for the common sense need to avoid dependence on a single supplier who might for example go bust). Later, two private sector companies[11] were selected to transport humans to and from the ISS, although the first flights under those contracts have not yet taken place.

So how will the private sector be involved in human exploration of the Moon? Bridenstine has spoken of using "public funds to support private equity and private funds to deliver more commerce, more economic growth, and solidify American leadership in space, science and discovery." Referring to industry supplying cargo and transporting humans to the ISS he indicated that this "can be extended to and around the Moon" (and even to Mars) [13] (see Fig. 9.5).

The first solid example of this new approach came in September 2018 with NASA requesting bids from industry to supply the initial part of the Lunar Gateway, a power and propulsion element, which as it names suggests will provide electric power (50 kilowatts) for the gateway and rocket propulsion to move it around. Later modules would hook up to it in much the way that the ISS was built up from separate modules. Two companies will be selected to build and launch a power and propulsion element, and then get it out close to the Moon. Only at the end of a one-year demonstration period will NASA decide if it will purchase either or both of the elements, which is around the time NASA anticipates launching additional components of the gateway. This unusual way to buy a spacecraft builds on industry's experience of deploying commercial satellites in a variety of orbits [14].

Fig. 9.5a and b. *Left:* The key official in delivering NASA's human spaceflight policy in the Trump administration is William H. Gerstenmaier, including the crucial issue of the role of industry. (Illustration courtesy of NASA.) *Right:* Former Congressman Jim Bridenstine at the podium just after being sworn in as NASA administrator by Vice President Mike Pence (at rear). Bridenstine has shown a willingness to seek increased roles for industry in the Lunar Gateway initiative, supported by Pence who chairs the influential National Space Council. (Illustration courtesy of NASA/Bill Ingalls.)

---

[10] Orbital ATK (previously called Orbital Sciences) and SpaceX.
[11] Boeing and SpaceX.

The big new factor compared to 50 years ago, or even 20 years ago, is the emergence of private sector companies willing to invest their own money to develop large rockets. We still have the old fashioned formula of NASA developing a new rocket with support from one of the aerospace industry giants – in this instance Boeing contracted to develop engines for NASA's Space Launch System (SLS), which will eventually be able to launch 100 tons or more into orbit. But we also have Blue Origin developing New Glenn, advertised to carry 45 tons to orbit starting in 2020 and SpaceX's Falcon Heavy, already capable of lofting 60+ tons into orbit (and plans for the much more powerful Super Heavy/Starship concept coming along). Given that NASA is committed to developing its own very expensive Space Launch System super-rocket, will it be able to find a way to use the much cheaper but similar (or better) rockets being developed by SpaceX and Blue Origin?

While NASA has to persuade Congress to approve its budget each year, both these private companies are owned by individuals with the personal wealth to make long-term investment decisions.

In the case of Blue Origin, owner Jeff Bezos is the world's richest man (his day job is running another of his companies, Amazon) with a vision of "a future where industry is moved off-Earth to take advantage of the energy and other resources of the Solar System." Shorter term, he says that "Today, we must go back to the Moon, and this time to stay", referring to *Apollo 11* as an event "out of sequence" and thus unsustainable [15]. He has indicated an interest in working with NASA to deliver cargo to the Moon's surface. If NASA didn't want to collaborate, Bezos said that "We'll do it. But we could do it a lot faster if there were a partnership [with NASA]."

SpaceX boss Elon Musk is also a billionaire[12] and thus able to pursue *his* visionary objectives. He developed the world's most powerful launcher, the Falcon Heavy, without the help of government funding. It is able to deliver more than 60 tons to Earth orbit, and within five months of its successful test flight in February 2018 SpaceX has signed up five paying customers. Musk was expected to continue development of Falcon Heavy so as to make it safe enough to carry astronauts into space, but instead will develop the much larger Super Heavy/Starship launcher able to carry about 150 tons into Earth orbit, intended for both cargo and human passengers – and reusable.

Will competition between SpaceX and Blue Origin spark a race back to the Moon? Jeff Bezos said that he was determined to achieve his vision of sustained industrial activity beyond Earth and would keep pursuing that vision until "I run out of money." He is apparently spending about $1 billion of his own money every year to fund Blue Origin, but with about $130 billion[13] to his name there's no

---

[12] His position on the Forbes "Rich List" fluctuates, as the value of Tesla and his other companies change on the stock market: on May 10, 2019, he was number 40 on the list.

[13] See for example [16].

immediate danger of that happening! We will return to the NASA, SpaceX and Blue Origin stories in Chapter 10.

The Military Aren't Interested

The military have been involved in space since the beginning. In the 1930s the young Wernher von Braun's interest in rocketry became serious when the German Army funded his group. The army had taken advantage of a loophole in the Versailles Treaty that ended World War I, which disallowed Germany from developing new artillery guns. Rockets weren't mentioned in the Versailles Treaty, so the army was willing to fund von Braun's developments in the hopes of getting a new form of artillery despite the poor reliability of the early rockets.

The V2 rocket, which von Braun eventually brought into production, was the first practical device capable of lofting objects into space – not yet into orbit, but above the atmosphere for a few minutes.

One of the ironies of von Braun's German military activities was that by working for the army he was restricted to unmanned vehicles. If he had been sponsored by the Luftwaffe (Air Force) he might have developed a rocket motor for what became the V1 cruise missile (nicknamed the "buzz bomb" by the Allies), which in turn might have led to something like the space shuttle rather than the Jupiter and Saturn rockets of the 1960s.

American space activities also began with the military. The early U. S. space programs under President Dwight D. (Ike) Eisenhower were started for military purposes. To avoid diplomatic incidents Ike authorized a civilian scientific space program to hide the parallel military surveillance satellite program ([6], p. 64). However, after the Soviets launched *Sputnik-1* in October 1957 Eisenhower did agree to a dedicated civilian space program, and created NASA to manage it. It made sense for many of the technologies used by NASA to be derived from or to proceed in collaboration with developments for the space programs of the Department of Defense. One early supporter of the Apollo program under President Kennedy was Secretary of Defense Robert McNamara. He welcomed the NASA funding that would go to the same aerospace companies that worked for his department, and would therefore become more effective without him having to pay for it.

Since the 1960s, the U. S. government investment in rockets has typically been authorized when both civilian and military spacecraft can use them. The same is true in other countries. In Europe for example the Ariane rocket was justified in the 1970s on the grounds that Europe wanted to launch its own military satellites without having to rely on the United States or the Soviet Union. However, it was realized that a rocket optimized for military purposes would be too expensive so it was decided to design Ariane to compete for commercial spacecraft contracts as the only way to make it affordable.

138 Why Is It Taking So Long to Return to the Moon?

In China most space programs are still managed by the military, including most launcher efforts. Excluding a few exported commercially to other countries, in recent years most Chinese spacecraft have been either purely military in function or with dual civil-military functionality. For example in 2011, of sixteen launches for domestic Chinese purposes six carried military satellites and seven carried dual use satellites [17]. In addition, and perhaps because of its use of advanced technologies, China's human spaceflight program comes under the military.

Back in the 1960s the U. S. military funded a human spaceflight program, but terminated it in 1969, just a month before the *Apollo 11* mission. The Manned Orbiting Laboratory (MOL) as the U. S. Air Force program was known was announced publicly to be investigating the military usefulness of placing man into space. Recently declassified information [18] has revealed that the actual mission was to place a manned surveillance and policing satellite in orbit.

Fig. 9.6. MOL Astronaut Group 1, selected for training in 1965. *L-R:* Michael J. Adams, USAF; Albert H. Crews, USAF; John L. Finley, USN; Richard E. Lawyer, USAF; Lachlan Macleay, USAF; Francis G. Neubeck, USAF; James M. Taylor, USAF; and Richard H. Truly, USN. (Illustration courtesy of the USAF.)

The thinking was that a manned reconnaissance system could more efficiently and quickly adjust coverage for crises and targets of opportunity than unmanned systems. With growing pressure on defense budgets due to the expansion of the

Vietnam War, the perceived duplication of effort with NASA programs and (perhaps most importantly) improved performance of unmanned surveillance systems, President Nixon pulled the plug on the MOL program. It had been underway for five years, spent $1.56 billion (about $8 billion in today's money) created a military astronaut corps (see Fig. 9.6) but never launched a manned spacecraft into space.

The U. S. military is interested in unmanned civilian space programs. Commercial and scientific spacecraft that provide telecommunications or surveillance are similar to those operated by the military. Generally speaking military unmanned spacecraft will have protection against hostile actions such as radiation from a nuclear explosion in space, radio jamming and blinding of cameras, but otherwise will use similar technology to commercial spacecraft and will be launched on the same rockets and from the same launch sites. The U. S. military is therefore keen to see commercial and scientific space programs grow, since they (the military) will benefit from advances in technology that emerge and from economies in sharing infrastructure such as launch sites.

The U. S. military actively supports the country's scientific activities in the Arctic and Antarctic, since both areas have strategic significance – especially the Arctic. Outer space is another area of strategic significance, and the military often collaborates with NASA in undertaking research into the space environment. However, the collaboration focuses on robotic spacecraft with little or no military involvement in human spaceflight.

When it comes to human spaceflight today, the U. S. military tend to see it as an activity that absorbs funding that could be better targeted at commercial or scientific spacecraft. Thus the U. S. military view of human spaceflight today is at best neutral and at worst may sometimes be opposed.

Turning now to Moon missions, the U. S. military did collaborate with NASA on one robotic mission to the Moon, the 1994 Clementine spacecraft that took what was then the most detailed images of the Moon's surface (see Fig. 9.7). Built by the Naval Research Laboratory, Clementine tested various new materials and equipment (the military's main interest) while also undertaking useful science (NASA's main interest).

If the Department of Defense *did* consider the Moon to be strategically important they would at the very least encourage NASA to establish a permanent presence there, and might even provide technical and financial support. However, inverting that argument, because the Moon is not a strategic area, NASA is on its own. Successive U. S. administrations have trumpeted their support for NASA to send humans back to the Moon but have avoided committing the necessary funds, implying that the goal is not one of strategic national importance.

If a human return to the Moon were perceived to be strategically important for the nation (as it was in the 1960s), the Department of Defense would be motivated to prevent other countries laying claim to it. Under those circumstances NASA might find enough political support to finance another Apollo.

Fig. 9.7. Thousands of images from the American Clementine spacecraft were combined to provide this image of the Moon's far side. The dark area in the upper left is Mare Moscoviense (Moscow Sea). Fig. 11.2 shows more detail of the lower half of the far side. (Illustration courtesy of NASA/JPL/USGS.)

Soviet (now almost exclusively Russian) space activities have been much less spectacular since the end of the Cold War, even under the aggressive President Vladimir Putin. The Russian military still launches satellites for surveillance, for secure communications and for navigation, but it has avoided big ticket items like the American eavesdropping satellites. Russian civilian science missions have also been relatively modest in recent years, far behind the ambitious U. S. missions to the giant planets, Jupiter and Saturn, and the rovers that have been exploring the surface of Mars for the past ten years. Having merged its space laboratory activities with those of the United States and its allies in the International Space Station, the Russians have to address what they will do once that station closes down. Russian military interest in human space missions has been very limited for the past 20 years. It is difficult therefore to see them supporting a human push out to the Moon.

For the past 20 years, China has been pursuing a series of five-year space programs. This included the launch of humans into space, then the launch of a laboratory in space visited by humans from time to time. The current plan is to extend that initiative and make the laboratory permanently or nearly permanently manned – perhaps with involvement of other countries. The plan also calls for development of new rockets, including eventually a very large rocket capable of placing more than 100 tons into orbit.

At the moment, China's manned space program is run by the military (as are about two-thirds of all of China's government space programs). The official rationale for military management of the program is one of convenience. The military has the skills and the facilities. Another rationale could be that the military wishes to influence and take ownership of any useful technology that emerges from the manned program. What is not clear is how the funding of the manned space program is approved in competition with demands for funds from other military projects – and there are plenty of those, from aircraft carriers to stealth aircraft to submarine-launched missiles and lots more.

What seems clear is if China publishes a long-term plan, there is top level approval for the developments within the plan. This approval is strongest for the five-year duration of each plan, and presumably weaker for developments identified as coming in later plans. Thus China is currently in the middle of an ambitious plan to explore the Moon using robotic probes. Objectives of future five-year plans are left vague but certainly do not rule out human exploration of the Moon.

Is the military interested in such a lunar program? The use of robots and drones by the military in war zones has rapidly increased in recent years, so we can understand military interest in the technology of robotically exploring a difficult to access and dangerous location such as the far side of the Moon (the current Chinese target). But sending humans seems to have little military rationale. What the Chinese military may be interested in is a rocket capable of placing large payloads in space. A human Moon program is an excellent disguise for developing such a rocket. We can't be sure what this Chinese large payload might be, but possibilities include a heavily defended spacecraft that might have defensive or offensive objectives.

The bottom line is that military interest in any particular space program guarantees lots of money for that program. The flip side of the coin is that with little or no military interest – such as to send humans to the Moon – the military will not help with the funding and may even lobby against it in order to prevent erosion of their funding.

References

1. https://en.wikipedia.org/wiki/Comparison_of_orbital_launch_systems accessed October 10, 2018.
2. https://history.nasa.gov/augustine/racfup2.htm accessed October 10, 2018.

3. Siddiqi, A., "Sputnik remembered: The first race to space," *The Space Review*, 2 Oct. 2017, accessible at http://www.thespacereview.com/article/3341/1 accessed October 10, 2018.
4. Siddiqi, A., *Sputnik and the Soviet Space Challenge*, University Press of Florida, 2003a.
5. Lafleur, C., "Costs of U. S. Piloted Programs," *The Space Review,* 8 March 2010, online at http://www.thespacereview.com/article/1579/1 accessed Oct. 10, 2018.
6. Norris, P., *Spies in the Sky*, Springer Praxis, (2007).
7. https://www.census.gov/foreign-trade/Press-Release/current_press_release/ft900.pdf accessed Oct. 10, 2018.
8. Jonathans Space Report No. 754, Oct. 8, 2018, online at http://www.planet4589.org/space/jsr/back/news.754.txt accessed Oct. 10, 2018.
9. Norris P, *Satellite Programs in the United States*, sections 43.6.1 and 43.6.2, in Schrogl K-U, et al (Eds.), *Handbook of Space Security*, Springer-Verlag (New York), 2015.
10. GSSAP Fact Sheet, online at https://www.afspc.af.mil/About-Us/Fact-Sheets/Article/730802/geosynchronous-space-situational-awareness-program-gssap/ accessed Oct. 10, 2018.
11. https://lunar.xprize.org/ accessed Oct. 10, 2018.
12. Klotz I, *Airbus Kicks Off New Moon Race*, Aerospace Daily & Defense Report, Oct. 1, 2018, online at http://aviationweek.com/space/airbus-kicks-new-moon-race accessed Oct. 10, 2018.
13. Foust J, *New NASA boss Jim Bridenstine faces his first challenge: a balancing act between the moon and* Mars, Space News, June 1, 2018, online at https://spacenews.com/new-nasa-boss-jim-bridenstine-faces-his-first-challenge-a-balancing-act-between-the-moon-and-mars/ accessed Oct. 10, 2018.
14. Foust J, *Where no commercial satellite bus has gone before*, Space News, 24 Sept. 2018, online at https://spacenews.com/where-no-commercial-satellite-bus-has-gone-before/ accessed Oct. 10, 2018.
15. Foust J, *Bezos and humanity's future beyond Earth*, The Space Review, June 4, 2018 online at http://www.thespacereview.com/article/3507/1 accessed Oct. 10, 2018.
16. Forbes Billionaire List: https://www.forbes.com/billionaires/#7b332472251c accessed May 10, 2019.
17. Norris P, "China aims for the high ground", *AEROSPACE*, Oct. 2012, pp22-25.
18. Day D, *The measure of a man: Evaluating the role of astronauts in the Manned Orbiting Laboratory program (part 3)*, The Space Review, http://www.thespacereview.com/article/3490/1 and links therein to Part 1 and Part 2 – accessed Oct. 10. 2018.

10

When Will the United States Go Back?

Three U. S. presidents (all Republican) have committed to sending humans back to the Moon and beyond: President George Bush, Sr., in 1989, President George W. Bush in 2004 (the Moon by 2020, Mars in the 2030s) and President Donald Trump in 2017 (the Moon by 2024, Mars in the mid-2030s). Brave words, but not accompanied by the necessary funding – so far.

To figure out when people are likely to return to the surface of the Moon, let's start by considering what NASA plans to do. We will then look at other U. S. initiatives from the private sector and then look beyond the United States, especially to China in Chapter 11 and then to Russia, as well as some supporting players such as Europe, Canada, Japan and India, in Chapter 12.

NASA

NASA chief Jim Bridenstine is changing the way that agency will go back to the Moon compared to what was done on Apollo 50 years ago.

- The first big change is that the landing on the surface of the Moon will be via a manned "Lunar Gateway" in orbit around the Moon, where astronauts will spend a year or more at a time (see Fig. 10.1). They will dispatch robotic probes to the surface and control them, and eventually there will be human trips down to the surface. The justification for putting a human space station so far from Earth is a bit vague except to argue that it would give useful experience for going to Mars. The first human landing in 2024 will use a very slimmed down version of the Gateway.

144 When Will the United States Go Back?

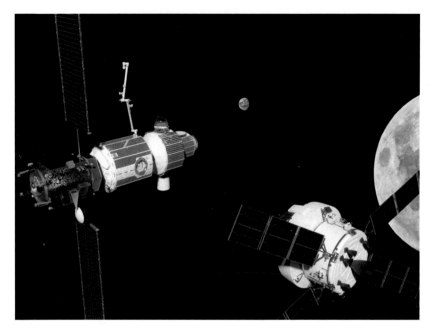

Fig. 10.1. Conceptual view of an early phase of the Lunar Gateway *(left)* being visited by a supply craft *(right).* (Illustration courtesy of NASA.)

- The second change is that international "partners" will be invited to participate. Apollo was totally an American adventure, but this time other nations can pay their share and get a piece of the action.

NASA has been spending billions of dollars each year for the past decade preparing for this Lunar Gateway initiative, so let's take a look at what has been going on.

The Rocket

NASA has been designing and funding a joint NASA/industry development of a giant launcher called the Space Launch System (see Fig. 10.2), just like it did in the Apollo days, but Bridenstine has been hinting that he will change that and buy commercial launchers instead. In particular he wants the launchers to be reusable, which the Space Launch System can't do. It's the old-fashioned type that ends up at the bottom of the ocean when its fuel is used up. In contrast, SpaceX and Blue Origin rockets return to base after launch to be refueled and used again (at least the large first stages of the rockets are re-used, but not (yet) the second or other stages).

When Will the United States Go Back? 145

Fig. 10.2. Artist's rendering of an early version of NASA's Space Launch System in flight. (Illustration courtesy of NASA/MSFC.)

The Spacecraft

NASA has also been funding industry to develop a Moon-landing spacecraft – the Orion. But again Bridenstine has been saying that he wants this craft to be reusable, which indicates that he's thinking of SpaceX's capsules that have been reused to service the International Space Station several times.

Industry's Role

NASA was planning to operate and provide the crew for the missions to the Moon, but as mentioned in Chapter 9, it already has changed things somewhat by asking industry to provide one of the major building blocks of the Lunar Gateway and get it out to its parking slot near the Moon. Contracts are to be awarded in 2019 for the module that will provide propulsion and electrical power for the gateway. The propulsion module will be able to alter the orbit of the Gateway. It could be distant from the Moon to ensure better communications with Earth or relatively close in to facilitate sending down a lander to the surface. Traditionally NASA would have said exactly what kind of propulsion and what form of electrical power it wanted, but this time around industry is left to work out the best way to provide these functions and to get the module out to the Moon in good working order.

Commenting on the most curious of the pieces of this jigsaw, NASA's Bridenstine explained the purpose of the gateway in orbit around the Moon as making it possible to "get to more parts of the Moon than we've ever been to before" – a clever play on the wording of a well-known beer advertisement, but without explaining how it will achieve this. Responding to criticism from astronauts and others, Bridenstine referred to President Trump's 2017 Space Policy Directive 1 that emphasizes sustainability and partnership (more than schedule). "The next time we go to the Moon, we're going to have American boots on the Moon with the American flag on their shoulders, and they're going to be standing side-by-side with our international partners who have never been to the Moon before," he said. "That's American leadership" [1].

Bridenstine also claimed that unlike Apollo NASA's approach would be sustainable, but he needs to change many of the details in NASA's plans or the reverse will happen.

Firstly, the giant Space Launch System (SLS) rocket that NASA is developing uses the 16 leftover engines from the space shuttle. These engines are well tested and have proven themselves to be safe – 100% success in 135 space shuttle flights. These Aerojet RS-25 engines have already been used for several shuttle flights, but in the SLS four will be used just once; use them, then lose them in the ocean. After four SLS launches there will be no more leftover shuttle engines. It's efficient to use leftover shuttle engines, but not exactly sustainable.

The plan is to re-start the production line of the space shuttle engines sometime in the 2020s under a $1.2 billion contract NASA has awarded Aerojet for six new RS-25 engines. The last engine was built about 20 years ago, so there may be some dust and rust to remove first. Assuming the production line can be got going again, each new engine will still be used just the once. As Bridenstine himself has said, the sustainable thing to do would be to reuse each engine on many SLS launches, as was done on the space shuttle, but instead each engine after being used once will end up in the ocean.

The RS-25 engines are powerful, but in fact in the first two minutes of each flight the majority of the SLS's thrust will come from two enormous booster rockets strapped to its side. These will be the most powerful rocket boosters in the world and are similar to those used on the space shuttle but about 25 percent more powerful. The experience built up in the space shuttle program is thus being used to advantage on SLS. We have watched videos of Elon Musk's Falcon rockets returning after a launch and landing upright, to be reused on a later Falcon launch, which seems like a sustainable way of doing things – see Fig. 10.5. Unfortunately the SLS boosters won't return to base after being used. Like fireworks they will burn all their powder and then fall back to Earth, ending up at the bottom of the Atlantic Ocean. Ironically, the Space Shuttle version of these boosters used to be recovered and re-used, giving us another example of SLS being at odds with Bridenstine's sustainability message.

The first stage of the SLS is powered by this combination of space shuttle engines and strap-on booster rockets. The second stage that carries SLS into orbit will use an existing design of engine at least to begin with – a United Launch Alliance rocket that powers the existing Delta IV launcher. This second stage is expendable, that is to say, it is discarded after use. The first SLS launch in 2020 will be a test run, carrying an empty Orion capsule out to the Moon, orbiting it for a week or so and then returning to Earth. The second SLS launch will be about three years later and will carry four astronauts out to the Moon, pass around it and then head back to Earth (without going into orbit around the Moon). These two missions are intended to show that SLS and Orion are fully operational for a range of deep space missions. Missions three and four will stick with the same type of second stage engine and will begin setting up the Lunar Gateway station in orbit around the Moon, and launch the lunar landing vehicle that will take American astronauts to the Moon by 2024 and home again.

From launch number 5 onwards, SLS will use four engines in the second stage instead of one; each of the four was expected to be an enhanced version of the engine family used in the first launch, but cost overruns of that design are causing NASA to consider using another engine. The new second stage will allow the SLS to carry a payload of 105 tons[1] into Earth orbit, compared to 70-95 tons for the less powerful first four launches. The first launch of this more powerful version of SLS is planned for about 2024. All of these second-stage rockets will be used on just one flight and then discarded.

SLS will eventually improve further and be capable of placing 150 tons or so into Earth orbit, thus exceeding the capability of the Saturn V, and in principle therefore able with a single launch to support a manned landing on the Moon's surface. The date for this super-advanced version of SLS to appear is a bit vague. "Late 2020s" was believable when the intermediate version of SLS (105-ton payload) was due in 2022, but since that date is now 2024, the 2030s seems more credible. This 150-ton payload version of SLS will require new and more powerful strap-on boosters – not a trivial exercise, since the boosters they are replacing are already "the most powerful rocket boosters in the world." The sustainable approach would surely follow the example of Elon Musk's Falcon rockets and return to be refueled and reused. Unfortunately that is not part of the Bridenstine NASA plan – the ocean floor beckons for each and every one.

Perhaps reusing the SLS engines in a sustainable way would have made its development too expensive. Building on space shuttle experience of strap-on boosters and using its old engines presumably kept development costs down. But

---

[1] There are at least three different weights that are pronounced "ton." For brevity, I use the word "ton" to signify a weight of 1,000 kilograms (about 2,205 lbs) instead of "tonne" or "metric ton." Note that in the United States and Canada, "ton" usually means 2,000 pounds, while in the rest of the world it usually means 2,240 pounds.

in fact the SLS development has not come cheap. By Sept. 30, 2018, NASA had spent $23 billion on the SLS, the Orion manned spacecraft it will take into space, and the ground facilities at Cape Canaveral and elsewhere, according to a 2017 NASA Inspector General report [2]. And NASA's 2018-19 budget continues that trend, providing more than $2 billion for SLS work, and a further $2 billion for work on Orion and the associated ground facilities.

Part of the reason for the high cost appears to be mismanagement by both NASA and its industry contractor, Boeing, according to an October 2018 report by NASA's own internal watchdog, the Inspector General. Boeing's costs are running at "double the amount initially planned," and the schedule "has slipped two and a half years and may slip further." The causes are bluntly identified: "Cost increases and schedule delays [are] largely driven by Boeing's poor performance" coupled with "poor contract management practices by NASA," including "overly generous award fees" paid to Boeing because of "flaws in NASA's evaluation of Boeing's performance [3]." You might think that heads would roll after such a blasting, but initially NASA's response had been to continue to support SLS (at least in public).

If Elon Musk's SpaceX and Jeff Bezos's Blue Origin can build similar rockets with little or no government development funds (as we will see below), does it make sense for NASA to spend $2 billion a year on SLS, especially when the industry contractor is doing a poor job? You can understand why Bridenstine is hinting at using commercial rockets, given the huge price savings that would be gained.

Longer term, NASA envisions funding regular missions to the Gateway and then to the Moon's surface by ending its $3-4 billion annual funding of the International Space Station (ISS) by 2025. The plan is for the private sector to step in and underwrite the operation of the ISS, with NASA just one of many customers buying services from the new owner as and when needed. There is considerable skepticism that the hugely expensive ISS will attract private sector investment on a scale that can replace NASA's deep taxpayer-funded purse.

Another 2017 NASA Inspector General report [4] said NASA's analysis "includes several overly optimistic assumptions related to revenues and costs," such as assuming that cargo transportation costs $20,000 a kilogram, while the current costs are about three times that. And the "scant commercial interest shown in the station over nearly 20 years of operation" led the Inspector General to question NASA's timetable.

In the commercial world you would just stop funding the ISS in 2026 and let it take its chances. However, it will be difficult to do that because of political support for its continued operation in the U. S. Congress and among partner countries, including Canada, Europe, Japan and Russia. Already in Congress draft bills are being introduced to extend U. S. funding of the ISS to 2030, supported by a mix of Republicans and Democrats. If the station is still subsidized by NASA to the tune of $3 billion or more a year after 2024, sending astronauts into orbit around the Moon will probably have to wait, and the date for regularly sending them to the surface of the Moon will slip further into the 2030s. Will Congress continue to pay to send Americans back to the Moon when the end date keeps slipping to the right? Some would say that was not politically sustainable.

Many aspects of NASA's stated plan are pretty shaky, unless Bridenstine can get a grip on the sky-high cost of the Space Launch System and its Orion capsule – by canceling them, for example. The leftover space shuttle engines are enough for four launches, two of which might be test flights and the next two could perhaps be used for deep space or other scientific robotic missions. Commercial launchers could then be used for transporting astronauts to the Lunar Gateway, eliminating the need for further investment in the Space Launch System.

Although the idea of the Lunar Gateway hasn't been justified very convincingly, let's not dismiss it out of hand. It has some interesting features that perhaps point to the way ahead.

The general idea for any attempt to send humans back to the Moon is to send unmanned robotic spacecraft there first. The robotic probes can orbit the Moon in order to provide communication and navigation links to the far side, which is always out of view of Earth. (China sent one such communications craft into position near the Moon in June 2018.) These orbiting probes can also provide detailed images of potential landing areas. Robotic probes can land on the Moon's surface, analyze its makeup, undertake experiments such as extracting oxygen from lunar rock, move around the area in rovers and even bring samples back to Earth. Only when these preparatory scouting missions have been completed do you send humans. All of these things were done by the United States and/or the Soviet Union in the 1960s and 1970s.

The Lunar Gateway is where humans could monitor all the robotic activity down on the Moon's surface, controlling events via radio links like you would a radio-controlled drone or boat. Astronauts on the ISS orbiting Earth have already remotely controlled a rover on Earth's surface, so controlling a lunar rover from orbit around the Moon should be equally feasible.

But back in the 1970s didn't the Soviets control the Lunokhod rover on the Moon from a control center on Earth? And China did the same in 2013/2014 with its *Chang'e 3* lander and Yutu rover and again in 2019 with *Chang'e 4* and Yutu-2 (see Chapter 11). Why do you now need to go to a Lunar Gateway to do it? There are some difficulties controlling a probe on the Moon by radio link from Earth, in particular the delay of at least one and a quarter seconds between the probe transmitting a signal from the Moon and you receiving it on Earth. This is just a fact of life, given that radio signals travel at the speed of light and the Moon is a quarter of a million miles away.

Now imagine you are steering a lunar rover from Earth, watching the TV images it sends back and turning the steering wheel by radio link to avoid obstructions and to get where you want to be. It will take one and a quarter seconds for the TV picture to reach you and another one and a quarter seconds for your radio signal to get back to the rover with turning instructions. So the action is delayed by two and a half seconds. And there will be other delays to the signals being routed to you from the radio aerial that is pointing at the Moon from some remote area hopefully not too far away from you – remote so that it is free from radio interference. So it is probably going to be three or four seconds delay instead of two and a half seconds.

You would clearly have to drive more slowly than on the open highway. You must see far enough ahead so that the rover won't meet the next boulder or pothole for at least three seconds.

If you were in the Lunar Gateway in orbit around the Moon, you would be only a few thousand miles at most from the rover compared to hundreds of thousands to Earth. So the delay is less than a tenth of a second instead of three seconds. You could drive the rover as if you were actually at the wheel.

You also want to collect samples and drill into the Moon's surface, which means you need to be able to send very precise instructions to the robotic equipment. For some tasks a delay of three seconds won't matter, but there might be some actions that need almost instant reactions. Operating the radio-controlled robots from a Lunar Gateway would make this possible.

Thus runs the argument in favor of having astronauts in orbit around the Moon to make exploration more effective and efficient.

But haven't they heard of self-driving vehicles, and of automated drilling and surgical equipment? With a self-driving vehicle (especially as there are no other vehicles or pedestrians or cyclists to worry about) you tell it the destination, and then it's hands-free driving until you're there. And the latest surgical tools can be programmed by the surgeon so that the machine actually drills into the patient, avoiding the risk of the surgeon's shaky hand.

Why not first try operating the lunar rovers and sample collecting machines from Earth? If the delay in the radio signals turns out to be a show stopper, then send astronauts into orbit around the Moon to reduce the delay. Surely the trends in robotic technology are all in favor of less rather than more direct human involvement.

Astronauts are already operating precision machinery remotely. In Fig. 10.3 NASA astronaut Jack Fischer is operating the "Justin" rover in Germany from the International Space Station in orbit 250 miles (400 km) above Earth.

Another set of experiments involves operating Canada's "Juno" rover in a quarry in St. Alphonse de Granby, Quebec, which has been selected because of its rugged Moon-like geology. Engineers at ESA's mission control in Germany are controlling the vehicle about 3,700 miles (6,000 km) away – see Fig. 10.4.

Fig. 10.3. NASA astronaut Jack Fischer operating a rover in Germany from the ISS August 28, 2017. Part of an ESA project, the experiment with the German space agency's robot, nicknamed Rollin' Justin, is about developing ways to allow astronauts to control robots from orbit. Fellow astronauts on the station Paolo Nespoli (Europe) and Randy Bresnik (NASA) also helped out. (Illustration courtesy of NASA/ESA.)

To check if precision operations can also be performed in this way, in another experiment European astronaut Andreas Mogensen placed a metal connector in a receptacle that had a mechanical tolerance to the connector of only six thousandths of an inch (0.15 mm). He performed this task in September 2015 not from the vicinity, but remotely – from the ISS, a signal traveling a distance of about 100,000 miles (160,000 km). The signal goes from the station to a geostationary relay satellite at an altitude of 22,000 miles (36,000 km) to a ground terminal in the United States, then through another geostationary satellite to Germany! Based on these and similar results, the operators are confident that humans could operate robots from Earth on the far side of the Moon – setting up a telescope there, for example [5][2].

But some people argue that a human presence on the Moon is needed. One British planetary scientist commented that "It will be much easier to drive a robot rover from a height of only a few hundred miles above the Moon than from Earth,

---

[2] http://esa-telerobotics.net/news/22/67/We-did-it-Sub-millimeter-task-from-Space

Fig. 10.4. October 2017, an ESA operator in Germany is controlling Canada's Juno rover in a quarry in St. Alphonse de Granby, Quebec, which has been selected because of its Moon-like features. Europe and the Canadian Space Agency are probing how to explore the Moon with a robot rover, investigating the challenges of remotely operating a rover in a representative lunar scenario with teams in several locations. (Illustration courtesy of ESA.)

which is 250,000 miles away; if you are on Earth, the control of a lunar rover [will be] much, much more difficult [6]." It will be interesting to see if China's remote operation of a rover on the hidden far side of the Moon (see Chapter 11) bears out these pessimistic remarks or confirms the optimistic results from Germany.

Now, it's clear that if we were talking about Mars and not the Moon the situation would be very different. The radio delay between Earth and Mars is at least 3 minutes each way and can be as much as 20 minutes each way. Astronauts in orbit around Mars would be able to take control of robots on the Mars surface, while from Earth you have to resign yourself to a long wait between sending a command and seeing the result. So a Mars gateway makes some sense, but a Lunar Gateway doesn't stack up.

Operating equipment on the Moon's surface isn't the only rationale for the Lunar Gateway. Another purpose is for astronauts to get used to living beyond the shelter of Earth's magnetic field (the ISS at 200-300 miles altitude is magnetically sheltered) and for NASA to figure out how best to manage a mission where the astronauts cannot easily be rescued. Before going all the way to Mars, the thinking goes, let's try it a bit closer to home.

For instance, the ISS recycles 90 percent of its water and 47 percent of its oxygen, but improved technology is needed to get these figure up to 98 percent (water) and beyond 75 percent (oxygen) if you're on a three-year mission to Mars [7]. The radiation hazards beyond Earth orbit require special protective measures and technologies. NASA administrator Bridenstine noted that humans lose 1-3 percent of their bone mass every month in space, so a trip to Mars (three years) is tough. He argues that the Moon allows us to test the technologies to address these issues and to learn how to mitigate problems as they occur.

A year-long mission in orbit around the Moon will give NASA and its partners confidence that a human mission to Mars can be safely undertaken. "The Moon is the best proving ground for going to Mars," said Bridenstine.

What hasn't been said is that you could get many of these results by having the astronauts stay in orbit around Earth but at a higher altitude – 22,000 miles up in geostationary orbit, for example. The radiation environment in geostationary is not as unfriendly as at the Moon, so it might be worth considering an even higher orbit at about 40,000-50,000 miles (64,000-80,000 km) altitude, say, in which the spacecraft would be outside the main sector of Earth's protective magnetic field for several hours a day. By avoiding going into orbit around the Moon (as proposed for the Lunar Gateway), this would be a lot cheaper than going out to the Moon. Not as exciting perhaps, but easier to access and monitor from Earth.

Some of the criticism of NASA's plans came in November 2018 from those you would expect to be its biggest fans, and in public. Even if you believe the official 2028 date by which NASA reckons it will regularly get humans back to the Moon's surface (and we saw above that "early 2030s" is more likely), "that comes across as having no sense of urgency" said the second to last man to set foot on the Moon, *Apollo 17* astronaut Harrison Schmidt. "2028 is so far off" said astronaut Eileen Collins (commander of space shuttle missions in 1999 and 2005), adding that "we can do it sooner." Former NASA Chief Mike Griffin said that a 2028 human return to the Moon was "so late as not even worthy to be on the table." Griffin went on to criticize the whole concept of the Lunar Gateway, calling it "a stupid architecture." Griffin isn't just some has-been former NASA boss (2005-2009); he is currently the Chief Technology Officer in the U. S. Department of Defense, and thus well placed to understand the technical issues involved.

Stung by these criticisms Vice President Pence announced in April 2019 that NASA would send humans to land on the Moon by 2024, warning that if NASA can't achieve that "we need to change the organization, not the mission." The space agency must transform into a leaner, more accountable and more agile organization, and must adopt an "all-hands-on-deck approach," he said [22]. Elaborating on this, Bridenstine explained that the third flight of the SLS would launch the Moon landing vehicle into space, and that a very early variant of the Gateway would be part of the plan. Experts outside NASA were quick to suggest schemes involving a few launches of SpaceX's already proven Falcon Heavy costing about $90 million each instead of NASA's planned use of the unproven SLS at more than $2 billion per launch [e.g.: 20].

The challenge now is for NASA to persuade Congress to fund the accelerated program. Outside experts have suggested that total extra funding of $15-25 billion will be needed, which Congress will find difficult to accept. The initial NASA request is for an extra $1.6 billion for the 2019/2020 financial year, with larger sums to be requested in the following years. NASA's Bridenstine describes this as a "down payment" and tries to sweeten the pill by promising that the first Moon landing mission will include a female astronaut. He also has named the initiative Artemis who in mythology was the twin sister of Apollo. [23, 24, 25] The tactic of asking Congress for less funding than expected at the beginning may well get the program underway, and give NASA time to work out the best technical approach to achieving a Moon landing by 2024. Some commentators have noted that in mythology, Artemis kills Orion (at least in some versions [26]), suggesting that replacement of the late and over-budget Orion spacecraft with a commercial alternative might be part of NASA's thinking.

Perhaps most damning is the view of the most senior of the surviving Moonwalkers, Buzz Aldrin. "I'm opposed to the Gateway," said Aldrin. "Why would you want to send a crew to an intermediate point in space, pick up a lander there and go down?" The Lunar Gateway as a staging point to the Moon's surface is "absurd" he said [8].

In summary, a quick look at NASA's plans is not very reassuring although better than when President Trump took up office. It looks like it will be the 2030s before NASA can get humans to the Moon on a regular basis using its own rocket. There is however now a chance that NASA will mount a flag-flying stop-gap mission in 2024 which would silence many of its critics. It risks being a one-off spectacular somewhat like the "space firsts" that the Soviet Union used to pull off in the early 1960s which made the headlines but diverted resources from long term goals. The SLS may not be ready for that date and would in any case be extremely expensive, so will NASA swallow its pride and use commercial launchers instead? Longer term the plans to send astronauts for long duration stays at a Lunar Gateway seem out of touch with the evolving technology of exploration.

That's the government way of doing it, so let's now see what the private sector has to offer.

SpaceX and Elon Musk

Elon Musk could have retired a very wealthy 31-year-old man back in 2002. He had been one of the key developers of the PayPal online payment system that eBay bought out for $1.5 billion that year. As the largest shareholder in PayPal, the payoff from eBay meant that Musk didn't need to work for a living any more. He decided to turn his attention to a boyhood dream of building rockets. That's how SpaceX took off.

Starting from scratch in 2002, Musk drew in experienced engineers from the Californian aerospace community to work on his vision of rockets that would return to Earth to be reused. He reckoned that the high price of rockets would stay high until you stopped throwing them away after just one launch. After a shaky start he has achieved his dream, and SpaceX is now the world's leading rocket manufacturer. In 2018 SpaceX rockets represented 18 percent of all launches worldwide and 68 percent of all American launches. After sending the second stage on its way into orbit, the first stage of a SpaceX Falcon 9 restarts three of its nine engines and brings the first stage back to Earth so that it can be reused a second time (see Fig. 10.5). In May 2018 a new version of the engine was introduced ("Block 5" is its official name) that can be reused up to ten times.

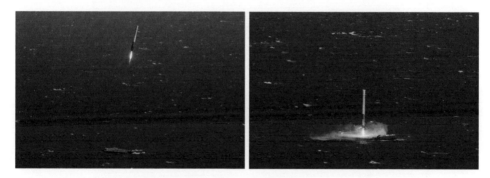

Fig. 10.5. The first stage of a Falcon 9 returns from space to land on a platform in the Atlantic Ocean: (a) approaching *(left)*, (b) touch down *(right)*. (Illustration courtesy of SpaceX. Used with permission.)

Musk is a Silicon Valley veteran, eating, sleeping and breathing the "anything is possible" and "work till you drop" ethic of the computer industry there (see Fig. 10.6). It is no surprise therefore that the exploits of SpaceX are publicized on social media, with videos on Facebook of the returning rockets landing on a barge at sea or at Cape Canaveral, and frequent tweets by Musk himself about each launch and other matters. The Silicon Valley culture extends to doing everything yourself as far as possible rather than subcontracting work to outside specialist companies. It also extends to being pragmatic and not just going by the book. The day before the second launch of the Falcon 9 in 2010, a crack was discovered in the nozzle of the second stage engine. Instead of spending a month taking the rocket apart and replacing the engine, Musk suggested that they simply cut the bottom off the nozzle just above the crack. His team agreed that the only impact would be a slightly less powerful engine, which they could live with. The half hour discussion led to a summons for a SpaceX technician to come to the launch pad armed with a pair of

156 **When Will the United States Go Back?**

Fig. 10.6. Elon Musk in 2015. (Illustration courtesy of Steve Jurvetson. Used with permission.)

shears. The next day Falcon 9 worked fine carrying the Dragon capsule into space on its first test flight. The 'can do' pragmatism and openness displayed by SpaceX reminds many people of NASA in the days of the Apollo missions. Unfortunately the average age of NASA staff rose after Apollo, reaching 50 in the 1990s, and NASA lost some of its youthful adventurous spirit ([9], pp. 121, 143, 176, 227).

Musk has used these principles in building his rockets and thereby driven the price down. His Falcon 9 is priced 40-50 percent below the sticker price of his main Western competitors such as Europe's Ariane 5. (Chinese prices are harder to compare because most of their business is for their own government; Russian prices also seem to be significantly higher than those of SpaceX.)

The low prices, impressive success rate (two failures in 72 launches as of May 11, 2019), primetime TV videos of rockets returning to the launch pad and in-your-face publicity has made the world's other rocket makers sit up and take notice. The Europeans have been forced back to the drawing board and plan now to phase out their expensive Ariane 5 and replace it with a smaller and cheaper Ariane 6. They have also forced two of their rocket companies to merge in an attempt to achieve economies of scale. They are investigating the possibility of reusing the engines but haven't gotten there yet.

You need a bigger rocket than Falcon 9 to go the Moon. In 2018 SpaceX carried out the first launch of the Falcon Heavy, capable of carrying more than 60 tons into Earth orbit. That first launch wasn't carrying anything useful, just one of Musk's Tesla electric cars as a publicity stunt. Yes, in his spare time Musk is also trying to turn the world of car manufacturing upside down by building the all-electric Tesla.

That Falcon Heavy was the most powerful rocket to leave Earth since the Saturn V did nearly fifty years ago. The next biggest rocket around at the moment – the Delta IV Heavy – can lift less than half the weight into orbit that Falcon Heavy can. Let's examine how SpaceX got funding to develop this super heavy lift rocket and compare it with what is happening with NASA's not-yet-launched Space Launch System mentioned above.

SpaceX started by investing its own money to develop a small rocket, the Falcon 1, which it marketed commercially, achieving the first successful launch in 2008. That first successful launch prevented the company going under, having followed several failed ones. Musk said he initially gave SpaceX a likelihood of success of just 10 percent. After one failed launch Musk noted that "failure sucks; there are a thousand ways a rocket can fail and only one that it can succeed."[3] "I wouldn't let my friends invest because I didn't want them to lose their money," he said of SpaceX. Instead he funneled his own money from the sale of PayPal into the business. "SpaceX is alive by the skin of its teeth. If things had gone a little bit the other way [it] would be dead [10, 11]."

Revenue and company funds were invested into developing the much bigger Falcon 9 rocket plus the Dragon spaceship that it could carry into orbit and that could return safely to Earth. At the same time NASA was seeking private companies to develop launchers and spacecraft that could carry cargo and eventually crew to and from the ISS. In August 2006, SpaceX was the first company selected in that program, winning a $276 million contract to complete the Falcon 9 development and that of the cargo version of the Dragon capsule. This led to the first successful Falcon 9 flights in 2010, including one that tested the Dragon capsule.

---

[3] The SpaceX team got to calling the explosion of one of their rockets "a rapid unscheduled disassembly," but that didn't stop the failures hurting.

158   When Will the United States Go Back?

The first delivery of cargo to the ISS followed in 2012 – "Looks like we've got us a dragon by the tail" declared astronaut Don Petit on the ISS.

SpaceX has since won more contracts from NASA as well as from the world's commercial space operators – for communications, navigation, surveillance and other satellites, and from the U. S. military. All of these wins have been in competition with other rocket manufacturers, so there has been no question of taxpayer subsidies. On the contrary, the entry of SpaceX into the market has pulled down the prices offered to the U. S. government by its competitors, typically by 20 percent or more, which has saved the U. S. taxpayer hundreds of millions of dollars (see Fig. 10.7). One dramatic example is the two contracts awarded by NASA in 2014 to Boeing and SpaceX to provide astronaut flights to the ISS. Boeing will

Fig. 10.7. Two pals out for a stroll. President Barack Obama *(left)* is escorted around the SpaceX launch pad by Elon Musk at Cape Canaveral in 2010. Obama's visit was part of his successful Congressional persuasion campaign to cancel NASA's Constellation super-launcher (its projected cost had doubled to $34 billion) and give the private sector a chance: "I think [Obama] wanted to get a sense if I was dependable or a little nuts" said Musk ([9], pp. 161-164). (Illustration courtesy of NASA.)

receive $4.2 billion for up to six operational flights each carrying four astronauts, SpaceX will receive up to $2.6 billion for the same number of flights ($1.6 billion less than Boeing). Thus SpaceX has saved the U. S. taxpayer $1.6 billion directly, and an unknown amount more if SpaceX had not been in the competition helping to drive down Boeing's price.

By 2017, SpaceX had also managed to fly the spent first-stage rockets back to a safe landing on Earth or onto a barge at sea. Videos of this uncanny feat, looking like something out of science fiction, went viral on YouTube.[4] The returned rockets were refurbished and used in future launches. Most Falcon 9 launches now use a rocket engine that has flown before, helping to drive down SpaceX's costs.

While winning business against traditional rocket suppliers, SpaceX began development in-house of the Falcon Heavy. The idea is simple – strap three Falcon 9s together and watch it fly (see Fig. 10.8). And recover the three rockets after the flight to be refueled and used again. Musk originally predicted that Falcon Heavy would fly in 2013, but it proved more difficult than he had hoped. Each Falcon 9 contains nine Merlin rocket engines burning kerosene and oxygen. Falcon Heavy

Fig. 10.8. First flight of the Falcon Heavy in February 2018. Essentially three Falcon 9 rockets locked together, it can place more than 60 tons of payload into Earth orbit, more than any other rocket since the Saturn V half a century before. Two of the three rocket cores made it back to their mid-ocean platform; the third tried but ran out of fuel. (Illustration courtesy of SpaceX. Used with permission.)

---

[4] See for example [12].

triples that to 27 Merlin engines, and requires complicated fuel transferring from the outside rockets to the central one. It turned out therefore that the central rocket (the "center core" is the SpaceX name for it) had to be extensively redeveloped in order to cope with the complexities. With fuel for the three rockets onboard "that's 4 million lb. of TNT equivalent," quipped Musk.

"It actually ended up being way harder to do Falcon Heavy than we thought," Musk told an audience in 2017. "At first it sounds real easy. You just stick two first stages on as strap-on boosters. How hard could that be? But then everything changes. All the loads change. Aerodynamics totally change. You've tripled the vibration and acoustics. The amount of load you're putting through that center core is crazy because you have two super-powerful boosters also shoving that center core, so we had to redesign the whole center core airframe. Then we've got the separation systems. It just ended up being way, way more difficult than we originally thought. We were pretty naive about that."

Around the time of the first launch in February 2018 he outlined some of the difficulties again. "Going through the sound barrier, you get these supersonic shockwaves. You could have some impingement, or when two shockwaves interact and amplify the effect and that could cause a structural failure as it goes transonic." Musk said. Engineers also were concerned about ice potentially falling off the upper stage and hitting the side boosters' nosecones. "That'd be like a cannonball coming through the nosecone," Musk said.

Despite extensive ground testing of the 27-engine configuration, "there's just so much that's really impossible to test on the ground" Musk said. One especially tricky problem involves aerodynamic forces between the bottom of the payload fairing and the side boosters. With the various pressure profiles and interactions from the side boosters, "it's a pretty busy area," SpaceX president and chief operating officer Gwynne Shotwell said.

However, according to Shotwell the most challenging part of the flight is probably during separation of the side boosters two and a half minutes after lift-off. Engineers had tested the separation mechanism and simulated the effects of boosters pushing away, but "you've basically got to go fly to execute those side-booster separations," said Shotwell in 2017. "That's going to be kind of a nerve-wracking moment for sure."

"The scary part of Falcon Heavy is all in the booster phase," she added. "The second stage is fundamentally the same [as Falcon 9]. The fairing is the same [the cover surrounding the payload at the top of the rocket]. So once we get through the first terrifying three minutes of the booster phase, then we're in much better understood territory."

Sure enough in the February 2018 launch, after two and a half minutes the side boosters separated, leaving a single-stick Falcon that burned for another 31 seconds. "It becomes like a Falcon 9 at that point," Musk said. The side boosters, both of which had been used on previous Falcon 9 missions, flipped around and headed back to landing pads at Cape Canaveral, touching down in unison accompanied by a quartet of sonic booms. Meanwhile, the central core separated from the upper stage and attempted to touch down on a drone ship floating about 300 miles (480 km) off Florida's east coast. The booster ran short of propellant, however, and was unable to restart its engines for a braking burn in the lower atmosphere, plowing into the ocean at 300 miles per hour [13].

The second stage of the Falcon Heavy continued into orbit, still carrying Musk's cherry-red Tesla roadster and later boosted itself out in the general direction of Mars.

More than 100,000 people turned up at Cape Canaveral and the nearby Florida coast to watch the historic Falcon Heavy launch. More than 2 million followed it live on YouTube, and a further 15 million watched it over the next two days, generating more than 30,000 comments. Eleven million also watched the views generated by the camera onboard the Tesla in space [14].

"Our investment to date is probably a lot more than I'd like to admit," Musk said at the time of the first launch in February 2018, later guessing the total to be "over half a billion, probably more."

Overall, the company anticipates a market for about four Falcon Heavy flights per year – two to fly national security payloads and two for commercial communications satellites. No customer has come close to requesting Falcon Heavy's full performance, which the company's website currently pegs at 140,660 lb. (63.8 tons) to low Earth orbit.

Let's just contrast the Falcon Heavy story with that of NASA's Space Launch System. Both began serious development in about 2010. By 2018 Falcon Heavy has flown successfully and is capable of placing 64 tons in orbit having cost perhaps $500 million of SpaceX funds to develop. The Space Launch System is hoping to fly for the first time in 2020, having absorbed more than 12 billion taxpayer dollars (24 times as much as Falcon Heavy) and capable of lifting 75 or 90 tons into orbit (the published figure changes from time to time). Falcon Heavy sells for about $90 million per flight while the Space Launch System will cost more than $1 billion per flight.

Instead of paying for one Space Launch System launch ($1+ billion), why not split your payload into two, launch it on two Falcon Heavy rockets ($180 million) and put the two bits back together again in space? OK it will cost something to do the splitting in two and the putting back together, but you will have $820+ million in the bank to play with.

One of the reasons Falcon Heavy is cheaper than the Space Launch System is that its technology is in some ways less advanced. The Space Launch System uses liquid hydrogen in its first stage (those leftover space shuttle engines), which has greater lifting power than any other fuel, but it has to be cooled to within 20 degrees of absolute zero (that is to -420° F/ -250° C) and kept at that temperature until used. The refrigeration machinery required for this is awesomely complex. Falcon Heavy uses kerosene, admittedly a specially refined form of kerosene called RP-1, which is similar to aviation fuel. You probably have a similar fuel in the tank of your car and perhaps a spare can of it in your garage. It doesn't give you the oomph per pound of liquid hydrogen, but it makes the machinery of your rocket an awful lot simpler.

Another Falcon feature is to use the same engine in the second stage as in the first stage – in both cases the Merlin engine. The second stage version has to be slightly modified to work in outer space, but its guts are the same. SpaceX builds Merlin engines – period. So the manufacturing process has been fine-tuned. The Space Launch System has a completely different engine in the second stage, and also uses two distinct engines in the first stage – the leftover shuttle engines and the giant strap-on boosters. So that makes three separate manufacturing processes to be mastered, and unfortunately the Space Launch System is the only launcher using two of them, so no chance to have fine-tuned their processes on another rocket.

The kerosene used by Falcon is pretty much the same fuel used to launch the first satellite, Sputnik, in 1957. Elon Musk realized that lowering the cost of rockets could be achieved not by using more sophisticated fuels but by applying common sense logic to their manufacture, including the idea of modular design – using the same part in different places. But having driven the manufacturing costs down Musk is now moving on to looking at new fuels and bigger engines.

The SpaceX plan had been to make steady improvements to the Falcon Heavy, for example to make it safe enough for astronauts to fly on. At the same time Musk planned to develop an even bigger rocket similar in performance to the Saturn V that took Apollo to the Moon.

In 2017 he announced that Falcon Heavy would not be developed further. It was available for launches of non-human payloads but would not be made astronaut friendly. He said that SpaceX has made good progress in designing and developing an even bigger rocket called Super Heavy, that will carry a massive spacecraft called Starship[5] and which is aiming for early "hopper" tests in 2019 (see Fig. 13.1).[6] This super-big rocket will use new engines that SpaceX has been developing with some funding (less than one-third of the total) provided by the U. S. Air Force. There will be 31 of the new engines in Super Heavy and seven in Starship. "This is a very big booster and ship," said Musk in 2017. "The lift-off thrust would be about twice that of a Saturn V. So it's capable of doing 150 metric

---

[5] The combined rocket and spacecraft previously had the working title Big Falcon Rocket or BFR.

[6] Hopper test: the rocket flies up a hundred miles or more and then returns to the ground a hundred or so miles away – it "hops" there.

tons to orbit *and* be fully reusable. So the expendable payload is about double that number." In other words it will have about twice the capability of the most advanced version of NASA's Space Launch System!

The new engine, called Raptor, has left kerosene behind and instead uses liquid methane – the main component of natural gas. It's much bigger than and about twice as powerful as the Merlin engine used in Falcon 9 and Falcon Heavy. The switch from kerosene to methane doesn't provide much more power,[7] but it burns more cleanly, avoiding the engine coking (clogging up with soot) and having to be cleaned before being used again.

Musk says they will reuse everything in the new rocket – not just the first stage (as for Falcon 9 and Falcon Heavy) but also the second stage and the fairing that protects the payload on top of the rocket. Musk reckons that the fairing represents about 10 percent of the cost of a Falcon 9 (i.e., about $6 million), hence the enthusiasm to recover and reuse it. He also wants to be able to turn around an engine after it lands much more quickly than now. Super Heavy and Starship "would enable reusability of both the boost phase and upper stage and fairing in a very high-throughput way. It is being designed for launch every few hours, whereas the Falcon architecture is designed to launch every few days [at best]" says Musk. So the clean-burning feature of methane that simplifies preparation for using the engine again is important.

All this rocket science disguises what Musk is trying to do. You will recall that the reason we don't send humans to the Moon any more is that the rocket you need is much too big and expensive to be affordable for anything else. Musk is turning this argument on its head. He is building a massive rocket and is going to use it for other space missions by making it cheaper to use than conventional throw-away rockets – cheaper even than Falcon 9 or Falcon Heavy. The cost of the fuel in a Falcon 9 is only one percent of the cost of the rocket, so if all you do is refuel the rocket and then launch again, the cost will be massively reduced. NASA tried this concept in the 1970s with the space shuttle, but in addition to topping up its tanks with fuel, each space shuttle needed $1 billion of work on it to fly again, becoming an expensive white elephant. Having made Falcon 9 and Falcon Heavy a success, few are betting against Musk delivering the vision with Super Heavy and Starship.

Another visionary feature of Musk's giant rocket is the choice of methane as the rocket fuel, because you could probably manufacture it on Mars. Yes, on Mars. Musk has said that he is driven to create big rockets so that humans can visit Mars – that's why he started SpaceX in the first place ([9], p. 238). He himself would like to end his life on Mars, "but not during the landing" he quipped. Manufacturing the fuel on Mars for a return journey to Earth would reduce the amount of fuel you have to carry with you from Earth. Robotic exploration of

---

[7] Internet chat forums are full of "on the one hand" methane is better in this way, but "on the other hand" kerosene is better in that way.

Mars has so far found no trace of oil, so getting kerosene there is out. But studies show that it should be possible to extract some simple chemicals from the Martian terrain, such as water, oxygen and, yes, methane.[8] If necessary, robotic probes could be landed in advance to begin the chemical processing and build up a supply in time for the arrival of a human crew.

Mars is important to Musk. "If there's a third world war we want to make sure there's enough of a seed of human civilization somewhere else to bring it back and shorten the length of the dark ages. It's important to get a self-sustaining base on Mars because it's far enough away from Earth that [in the event of a war] it's more likely to survive than a Moon base," Musk said. It will be very dangerous, so it's definitely not an escape hatch for rich people. "It will be like Shackleton's ad for Antarctic explorers: 'Difficult, dangerous, a good chance you'll die, excitement for those who survive,'" he said. Getting to Mars "will be the greatest adventure ever – in human history ([9], p. 123)."

So being able to make rocket fuel on Mars is an important issue for Musk, suggesting this could be a key reason for switching to methane. However, methane does pose some difficulties compared to kerosene. For one thing you have to cool it down to $-163\,°C$ ($-261\,°F$) to turn it from a gas to a liquid. This is similar to what you have to do to turn oxygen into a liquid that will be mixed with the methane to burn, which is already done in the Falcon rockets.

Whatever SpaceX's reason for switching from kerosene to methane,[9] by the early 2020s SpaceX will have a rocket comparable in performance to what NASA's Space Launch System hopes to achieve in the late 2020s (150 tons to orbit) and funded mainly by its own company funds.

So far, SpaceX has not carried astronauts on its rockets – although the Dragon capsule that brings cargo to and from the ISS has windows! It plans to start doing so in 2019 under its contract with NASA to ferry astronauts to and from the ISS. The mission will use a new version of its Dragon cargo ship called Crew Dragon launched on top of the latest version of the Falcon 9.

Astronauts will have to stick with Falcon 9 for the moment, since SpaceX has said it will not make the necessary upgrades to the Falcon Heavy to certify it for human spaceflight. So airline-style trips to the Moon and beyond with SpaceX must wait until Super Heavy and Starship are working and have been shown to be sufficiently reliable for astronaut use. Until then a human Moon landing mission would require perhaps two Falcon Heavy launches to place some provisions on the Moon's surface and carry a Moon lander into Earth orbit followed by a Falcon 9 launch to take the crew up to it – a scenario suggested by space guru Robert Zubrin ([20]).

---

[8] See for example [15].

[9] Two examples of such reasons among many: methane gives off less pollution than kerosene when burned; methane can be extracted from sewage gas and thus help meet recycling targets.

Will That Be Before NASA's Space Launch System?

Musk admits that his forecasts of delivery dates have often been wrong. "People have told me that my timelines have historically been optimistic," he said. Assuming successful "hopper" tests in 2019, orbital launches are expected to occur by 2020. So, Musk's earlier statements that the first flight of the Super Heavy/Starship would take place by 2022 and the first crewed flight by 2024 still appear to be on. The first crewed flight of the Space Launch System is also scheduled for 2024, so it's neck and neck at the moment, with delays likely for both of them.

In fact, in September 2018 SpaceX sold tickets for a 2023 Moon trip on the Super Heavy/Starship. And Musk added that he planned the first human trips to Mars the following year! The 2023 six-day Moon trip won't actually land on the Moon but will instead swing around it and come straight back – a circumlunar flight. Japanese fashion retail billionaire Yusaku Maezawa paid an undisclosed price to buy tickets. Musk said Maezawa had paid a significant deposit and would have a material impact on the cost of developing the new rocket, which he estimated at about $5 billion,[10] while Maezawa said that it cost more than a Basquiat painting. (He is famous for having paid $110 million for a Michel Basquiat painting in 2017.) That fee has purchased at least nine places on the flight – Maezawa accompanied by 8 yet-to-be-selected artists. The spacecraft can carry many more passengers, so NASA astronaut Scott Kelly was quick to tweet: "Yusaku Maezawa, this will be a great adventure! Good luck on your trip, and if you need someone with a little experience to go with you, my schedule is wide open in 2023."

There has to be some doubt that SpaceX can have the Super Heavy and Starship ready for human passengers by 2023. "The 2023 date is definitely not certain," admitted Musk. He went further and pointed out that the huge development challenge meant that "it's not 100 percent certain we can bring this to flight" at all.

Beyond a circumlunar flight, the SpaceX website outlines how the Super Heavy and Starship can undertake a Moon landing. It would involve two launches, with the second one acting as a refueling ship for the one that lands on the Moon. "If we do a high elliptic parking orbit for the ship and [refuel] in high elliptic orbit we can go all the way to the Moon and back" (See [16]).

While the ultra-big Super Heavy first stage seems like the most difficult challenge in all this, Musk begs to differ. Development of the first stage will be relatively straightforward because "it's like the Falcon 9 boost phase, but with 31 engines instead of nine. I don't want to get complacent, but I think we understand reusable boosters," says Musk. SpaceX will focus first on the spaceship section,

---

[10] Musk clarified the $5 billion estimate as "not more than $10 billion and not less than $2 billion" in the Sept. 17, 2018, press conference when announcing the identity of the first Super Heavy/Starship customer.

which Musk reckons is by far the more technically challenging part of the system, mostly because of extreme heating while re-entering Earth's atmosphere from the Moon or Mars transfer velocities of 25,000 mph or more. "Certain elements of re-entry heating scale to the eighth power," Musk says. "I didn't think there was anything that scales to the eighth power, so testing that ship out is the real tricky part." The Starship has an enormous 1,000 cubic meters/35,000 cubic feet of payload volume, with 80 percent of this pressurized for carrying passengers and crew – bigger than the main deck on the Airbus A380 (the world's largest airline).

As for going to Mars, the SpaceX vision goes as follows: the massive Super Heavy first-stage would carry aloft the giant Starship second stage. Once this combination is launched, Starship would detach and use its thrusters to assume a parking orbit around Earth. Super Heavy would then guide itself back to its launch pad, take on a propellant tanker, and return to orbit. The propellant tanker would then attach to the Starship and refuel it and return to Earth with Super Heavy. The Starship would then fire its thrusters again and make the journey to Mars with its payload and crew.

As one French commentator put it after the Falcon Heavy launch: "Elon Musk continues to build his Martian project and to inspire people to dream. Just like in the Apollo days [17]."

Blue Origin and Jeff Bezos

Elon Musk wants to go to Mars so that humanity will have a safe haven in case Earth is destroyed. Jeff Bezos is another space-obsessed billionaire (see Fig. 10.9), and he wants to go the Moon rather than Mars as a step towards a future where industry is moved off-Earth to take advantage of the energy and other resources of the Solar System. He envisions Earth being zoned for residential and light industry with heavy industry undertaken in space. The Blue Origin rocket company he founded to drive forward this vision takes its name from Earth as seen from deep space: a pale *blue* dot in an ocean of black and the *origin* where humanity began.

"Today, we must go back to the Moon, and this time to stay," he said in May 2018 after describing how *Apollo 11* was an event "pulled forward, out of sequence" and thus not sustainable. Permanent human presence on the Moon was essential to his long-term vision of millions of people living and working in space. "I always thought that this idea of going to Mars without building a permanent base on the Moon would end the same way Apollo did, where we would do it, there would be a ticker-tape parade, and then 50 years of nothing."

"We're so lucky to have the Moon. It's so conveniently located," he went on. "We now know things that we didn't know before. We know that there are

Fig. 10.9. Jeff Bezos is the CEO of Amazon and the world's richest man. He has been funding Blue Origin, his space rocket company, to the tune of $1 billion a year for several years. Here in 2015 he inspects the western Texas launch facility of the New Shepard rocket before its maiden voyage. (Illustration courtesy of Blue Origin. Used with permission.)

volatiles trapped in the dark craters of the Moon that are perpetually shaded. We know that there's water there. There's ice there. There are probably other interesting things in those craters as well." In previous public statements he had avoided choosing a preferred destination, but these more recent remarks make clear that establishing a human presence on the Moon was a priority for him.

The Bezos fascination with human visits to the Moon was strengthened by his successful expedition to raise the mighty Saturn V first stage engines from the bottom of the Atlantic Ocean, where they had lain for 50 years (see Fig. 2.2). He brought in a team with experience in recovering artifacts from the wreck of the Titanic,[11] and in 2011 they located a large number of objects that seemed likely to be parts of the Saturn V rockets – more than 1,500 feet/450 m deeper than the Titanic at a depth of about 14,000 feet/4,200 m. In 2013 Bezos and several members of his family joined the 50-man team on the three-week voyage to recover the objects. They used two remotely operated mini-submarines to haul up the objects

---

[11] The world's largest ocean liner that hit an iceberg in the north Atlantic in 1912 on its maiden voyage from Britain to New York, sinking with the loss of about 1,500 lives. The wreck was located in 1985 by oceanographer Robert Ballard. Later expeditions recovered artefacts from the wreck.

they found, hosed the mud off them and brought them back to dry land to see what they had got. A specialist museum in Kansas determined that the haul included the thrust chamber from the center of the five engines that powered *Apollo 11* on its way to the Moon back in July 1969. Bezos said that the recovery crew "had the feeling that we were recapturing history and making some history at the same time." His trademark booming laugh rang out as he added "I can tell you for sure we had a lot of fun doing it" ([9], pp. 189-197).

"You have to lower the cost of access to space to do these grand things that we're talking about," he said – and Elon Musk would surely agree with him on that. Bezos has been funding his Blue Origin rocket company to the tune of about $1 billion a year for the past few years in order to achieve this. He has yet to send anything into orbit, but he has built and demonstrated a sophisticated single-stage rocket that can return to Earth after launch and be reused a number of times. It is powered by an engine using liquid hydrogen and liquid oxygen – a powerful fuel combination but one that requires very complex engineering.

The rocket is called New Shepard (see Fig. 10.9), after America's first man in space, Alan Shepard, whose Mercury flight in 1961 was suborbital. It went up 100 miles and came back about 300 miles from the launch pad without going into orbit. New Shepard will provide a similar trip for paying customers starting in 2019 – the price had not been revealed as these pages are being written. The New Shepard space tourist will have a much better experience than Alan Shepard, whose spacecraft landed (intentionally) in the ocean. New Shepard passengers will return to dry land sitting comfortably in their capsule as it floats down under parachutes, the final drop cushioned by small rocket engines. The rocket part will return separately to its launch pad using its engine to settle down smoothly right where it had taken off.[12]

The experience will also avoid the intense g-forces that Alan endured, although there may be some discomfort as the capsule slows down during its return – less than half the 11-g forces that Alan sustained, Unlike Alan in 1961 the modern Shepard will have room to float around in zero gravity, experiencing the weightlessness of space for a few minutes and turning somersaults in the air if you wish. Finally, instead of Mercury's small porthole, New Shepard will have big picture windows through which the space tourist can see the blackness of space, the curve of Earth's horizon and the spectacular views of Earth below. Bezos says "I will definitely go. I can't wait, actually."

The New Shepard capsule can hold six people – all passengers as the flight will be controlled from the ground. It should be an exciting experience for them, not least if one or two of them suffer from space sickness. In zero gravity, vomit has

---

[12] Videos of this dramatic return from space of the rocket can be seen on YouTube; for example the July 2018 landing is 40 minutes into the video at https://www.youtube.com/watch?v=NRDhdHRyyjc&feature=youtu.be

an unfortunate habit of floating around, so let's hope the capsule carries a readily accessible vacuum cleaner to suck it up before it gets in the hair of the other passengers. The Blue Origin website doesn't mention this.

The up and down experience of New Shepard is just a small first step towards the Jeff Bezos vision of manufacturing on the Moon. The next step is to get into orbit, and for this Blue Origin is building the New Glenn rocket (see Fig. 10.10), named after the first American to orbit Earth, John Glenn (1962). Initially it will carry satellites into space, but the plan is for it to be certified for human spaceflight as quickly as possible.

Fig. 10.10. Artist's rendering of Blue Origin's New Glenn launcher. Initial launches from Spaceport Florida are planned for 2020. (Illustration courtesy of Blue Origin. Used with permission.)

New Glenn will be able to place 45 tons in orbit, second only to the Falcon Heavy in the list of world rockets (unless NASA's Space Launch System gets into orbit ahead of it). The current plan is for the first launch into orbit to be in 2020, although that will probably be just a test launch (without a real payload on top). Regular flights to orbit with commercial satellites would then begin in about 2022.

When Will the United States Go Back?

No reliable date is available for when astronauts will be lofted into space aboard New Glenn. Since the first test launch hasn't even taken place yet, a date in the mid-2020s for it to carry human passengers to orbit seems about the best we can hope for.

Jeff Bezos is not yet as widely known as you might expect for the world's richest man. He started Amazon in a garage in Seattle in 1994 and drove the idea of online shopping forward while building Amazon from the bottom up. He remains the chief executive officer and following his recent divorce owns 12 percent of the stock, which puts his wealth at about $120 billion. Compared to the impact Amazon has had on economies and shopping habits across the world, Jeff's sideline of building rockets and spaceships seems curious – but it's his money, so he's entitled.

Bezos' fascination with space goes back to July 20, 1969, when he was five years old and watching Neil Armstrong and Buzz Aldrin land on the Moon. "It really was a seminal moment for me," he said later. "I remember watching it on our living room TV. That definitely became a passion of mine." As valedictorian, his speech at high school graduation was about space – "space, the final frontier, meet me there" it concluded. Studying engineering at Princeton University his interest in space remained strong, becoming president of the Princeton chapter of a nationwide student organization called Students for the Exploration and Development of Space (SEDS). His former high school girl friend said that Jeff founded Amazon in order to make enough money to start a space company. He admits that there "is some truth to that" but perhaps mindful that such a remark could trigger an Amazon stock price nose dive, quickly adds that Amazon is his real passion and not a steppingstone to Blue Origin ([9], pp. 62-63, 67, 70, 254-255).

From the start of the Space Age in 1957 to the end of the 20th century, rockets were primarily developed by governments. As Elon Musk put it in 2008 when his Falcon 1 finally made it into orbit: "It's normally a country thing, not a company thing." Then, barely into the 21st century, two private sector entrepreneurs succeeded in entering the business.

Both Musk and Bezos have emphasized the importance of reusing the rockets instead of discarding them after a single flight. Musk's Falcon 9 has shown that this can be done, although the hoped-for drastic reduction in price has not yet appeared. (Until now, Falcon 9 has been cheaper than other rockets because it is manufactured efficiently not because it is reused.) The newest version of Falcon 9 is intended to be reused about a dozen times instead of a single reuse for earlier ones, so we may start to see steep price reductions soon.

Blue Origin, too, has emphasized reuse from the beginning, and the New Shepard rocket returns to the launch pad after each flight. The liquid hydrogen and liquid oxygen its engine uses puts it in the same category as NASA's Space Launch System and the space shuttle, but it is much smaller than the engines of those giant rockets. Nevertheless it is a major achievement for Blue Origin to have succeeded in developing such engines and being able to reuse them. The extreme low temperature needed to store the liquid hydrogen is an especially difficult challenge. Blue Origin reckons that a series of NASA contracts starting in 2010 and eventually valued at $25 million probably knocked a year off the time it took to develop the engine ([9], p. 180, see Fig. 10.11). Hydrogen is inherently a cleaner fuel to use than the kerosene-fueled engines of Falcon 9, which should simplify cleaning the engines after each flight.

Fig. 10.11. Jeff Bezos *(center-left)* with then NASA Deputy Administrator Lori Garver at the Blue Origin factory in Seattle in 2011 next to the crew capsule. (Illustration courtesy of NASA/Bill Ingalls.)

For the much bigger New Glenn, Blue Origin is adopting the methane fuel approach that the rival Falcon rockets are just beginning to use. There will be seven of these new Blue Origin engines in the first stage of each New Glenn rocket, and another rival rocket builder, United Launch Alliance, has decided to purchase the engine for its Vulcan rocket. The engine has been successfully test fired on the ground (Fig. 10.12) and seems on course to be ready for the first New

Fig. 10.12. Ground test of the BE-4 engine, seven of which will make up the first stage of the New Glenn rocket. (Illustration courtesy of Blue Origin. Used with permission.)

Glenn test flight in 2020. The second and optional third stages will use a variant of the New Shepard engine with its liquid hydrogen and liquid oxygen fuel mix.

Blue Origin Vice-President Clay Mowry explained the advantage of developing the suborbital New Shepard before New Glenn. "For us, New Shepard is really a vehicle we are using to teach ourselves how to launch, how to ramp up, how to refurbish the vehicle and to re-fly again, and to do that at a much lower cost than we can do with the orbital vehicle – about a 50th the cost of flying an orbital mission." He said that Bezos has invested $2.5 billion in New Glenn, and that the rocket has no funding from the U. S. government [18].

One of the differences between Bezos and Musk is the commercial approach of their ventures. Musk's SpaceX has tried to win commercial customers from the beginning, then using the revenues from those sales to fund bigger versions. This obviously reduced the amount of Musk's own funds that had to be sunk into the company – and at the start Musk didn't have the billions that Bezos has. Another advantage is that the customers tell you what *they* want, which helps to ensure your product is commercially viable.

The Bezos approach of funding most of the development himself and being quite secretive about it until recently is very different. The advantage is that he avoids having to diverge from his goals because a customer wants something different. And he avoids having to justify what he is doing to external shareholders and clients. The downside is that he has to provide most of the funding, but he can afford it. However he is now starting to sell his rockets commercially – four customers have signed up to launch satellites on New Glenn, all from the communications and broadcasting sector. Mowry said that New Glenn's fairing (the cover that protects the payload) was increased in size from 17¾ feet wide to 23 feet (5.4 m to 7 m) as a result of input from market demand and customer reactions, making it possible to launch more small satellites or geostationary satellites with larger antennas and structures. The sale of the new engine to United Launch Alliance announced in October 2018 is another example of Blue Origin getting seriously involved in the market.

New Glenn will be the second or third most capable rocket in the world when it becomes operational, but it is a long way from being able to carry passengers to the Moon's surface. Its payload capacity of 45 tons to Earth orbit is well short of the 150 tons or so you need for that. For that you probably need the "New Armstrong" rocket. Being named after the first man to walk on the Moon's surface is a bit of a giveaway as to its objective. Bezos said New Armstrong would be the next project after New Glenn but gave no details about what it is comprised of or when it would appear. In 2016 he teased journalists with the remark that "Up next on our drawing board: New Armstrong. But that's a story for the future [19]."

Further proof of Bezos' focus on the Moon came in May 2019 when he unveiled the Blue Moon spacecraft designed to land cargo and eventually people on the Moon [21]. The timing was probably aimed at impressing NASA in anticipation of its request for Moon lander designs consistent with the 2024 target set by Vice President Spence the previous month. "We can help meet that timeline" said Bezos, "because we started three years ago." The launcher to be used to propel Blue Moon to its destination wasn't mentioned by Bezos – presumably he will go with whatever NASA proposes.

As things stand, New Glenn is two or three years behind SpaceX's Falcon Heavy, with both in the 45-65 tons in Earth orbit category. Bezos's New Armstrong rocket and Musk's Super Heavy/Starship will both presumably be in the 150 tons to Earth orbit category. Unless things change it seems likely that the Bezos Moon rocket will be at least three years behind Musk's.

Bear in mind, though, that Musk is focused on getting to Mars much more than the Moon, while for Bezos the opposite is true. Will this influence the development of their respective giant rockets?

Musk and Bezos have found themselves in competition from time to time, and each time sparks have flown. In 2013 the two companies were the only bidders to take over launch pad 39A at Cape Canaveral – the now run-down launch pad where *Apollo 11* and all the other Saturn V flights had begun. NASA selected the SpaceX proposal, but Blue Origin filed a legal protest claiming that pad 39A should be made available to multiple launcher companies, not monopolized by one. Bezos ramped up the tension further by getting formal support from SpaceX's main rocket competitor, United Launch Alliance, and by lobbying hard in Congress. Musk was apparently infuriated by a company "that hadn't even gotten a toothpick into orbit." Things got tenser as Musk suspected that Blue Origin was poaching his staff. This reopened an old wound, because SpaceX had sued (unsuccessfully) a former employee back in 2008 for allegedly taking SpaceX proprietary information to Blue Origin.

Musk's irritation was all the greater because Blue Origin was only aiming for suborbital spaceflight. He threw the gauntlet down in an open email to the *Space News* trade journal, saying that if Blue Origin had a rocket available within five years capable of reaching the international Space Station then they could use Pad 39A. But "[F]rankly, I think we are more likely to find unicorns dancing in the flame duct." SpaceX was awarded Pad 39A, and Bezos didn't respond to Musk's childish "unicorns" jibe. The irony seemed to be lost on Musk that Blue Origin was taking on the role against the industry leader (SpaceX) that he, Musk, had taken on ten years earlier against the established space companies such as Lockheed. Blue Origin settled for taking over Pad 36 from which many important space missions had begun, but lacking the iconic status of Pad 39A a few miles away.

In 2014 Musk reacted incandescently to another Blue Origin initiative. The U. S. Patent Office approved Blue Origin's patent application to recover a launch vehicle on a seagoing platform. SpaceX filed suit to challenge the patent, pointing to the decades of prior work on such concepts (so-called "prior art" in the patent world) including a 1959 Russian science fiction movie showing a rocket landing on a ship at sea. The patent would have crippled SpaceX's plans for recovering its first stage rockets (see Fig. 10.5). Blue Origin eventually withdrew most of the claims in its patent, but Musk was still angry about the affair years later.

Musk's tendency to react immediately and publicly to perceived jibes from Bezos emerged again a year later. In November 2015, Blue Origin flew its New Shepard rocket to a height of 62 miles and then successfully returned it to the launch pad. The favorable attention this achievement attracted spurred Musk into social media action. "SpaceX Grasshopper rocket did 6 sub-orbital vertical takeoff and landing flights starting in 2013" he tweeted somewhat peevishly. To an outsider this seemed unfair because the Grasshopper had only risen at most 1,000 meters (3,280 feet) whereas New Shepard made it beyond the 62-mile (100-km)-high boundary of space.[13]

A month later SpaceX successfully landed the returning first stage of a Falcon 9 orbital mission at Cape Canaveral close to where it lifted off ten minutes earlier – the first time it had succeeded in doing this. Having resisted rising to Musk's jibes in the past, on this occasion Bezos decided to liven things up. He tweeted "congrats on landing Falcon's suborbital booster stage. Welcome to the club!"

---

[13] Some people put the boundary at an altitude of 50 miles.

SpaceX supporters responded aggressively on social media, and Musk felt that he didn't need to get personally involved. The fans reminded Bezos that the Falcon 9 went into orbit while New Shepard didn't. "Not even in the same league buddy" said one. In fact, Bezos had a point. The second stage of the Falcon 9 that actually goes into orbit has never been recovered – that would indeed have been a major step forward. The Falcon 9 first stage tops out at about 50 miles (80 km) while flying at about 3,700 mph (6,000 kmph), compared to New Shepard reaching 62 miles (100 km) and about 2,400 mph (3,900 kmph), in other words, a broadly similar challenge.

Blue Origin's New Glenn rocket will compete directly with SpaceX's Falcon 9 and Falcon Heavy, and indeed has already won a handful of contracts that might otherwise have gone to SpaceX. As mentioned above, Blue Origin will also supply the engine it has developed for New Glenn to the United Launch Alliance for its Vulcan rocket. Only in 2015 had SpaceX persuaded the Department of Defense to allow it to bid on launching military satellites – the monopoly of United Launch Alliance for a decade. SpaceX had then gone on to win almost all of the military contracts up for grabs, including five of the first six new generation GPS satnav satellites.[14] The Alliance had been forced to shut down its Delta and Atlas rocket series and to begin development of the new Vulcan rocket. By supplying a key engine to the Alliance, Blue Origin is setting its sights on SpaceX's main stream of business – a more serious competitive challenge than exchanges of tweets ([9], pp. 181-183, 200, 205, 221-229, 241-247).

In any case, the good news is that there are likely to be *two* American private sector rockets capable of repeating the Apollo story by the mid- to late-2020s. And they are fierce competitors, which augurs well for continued progress towards the goal of humans on the Moon (and even Mars). Meanwhile, NASA might even get the Space Launch System to the starting line too.

Now let's see if there is a non-American entrant in this new Moon race. First up in Chapter 11 the giant in the East: China. Then a quick check in Chapter 12 on Russia, if only because it was the only other competitor the last time this race was run, plus Europe, Japan, India and Canada.

References

1. Foust, J., "Is the Gateway the right way to the moon?", *Space News Magazine*, Dec. 17, 2018, pp11-14 online at https://bt.e-ditionsbyfry.com/publication/?i=551411#{"issue_id":551411,"page":0} (subscription required) accessed Jan. 12, 2019.
2. Clark, S., *NASA expects first Space Launch System flight to slip into 2020*, Spaceflight Now, Nov. 20, 2017, online at https://spaceflightnow.com/2017/11/20/nasa-expects-first-space-launch-system-flight-to-slip-into-2020/ accessed Oct. 10, 2018.

---

[14] The first of these was launched on a Falcon 9 on December 23, 2018.

3. NASA Office of Inspector General, NASA'S Management of the Space Launch System Stages Contract, Report No. IG-19-001, October 10, 2018, online at https://oig.nasa.gov/docs/IG-19-001.pdf accessed May 13, 2019, pp 3–4.
4. NASA Office of Inspector General, *NASA's Management and Utilization of the International Space Station*, July 30, 2018, Report No. IG-18-021, online at https://oig.nasa.gov/docs/IG-18-021.pdf accessed Oct. 10, 2018.
5. *Orbital 'peg in the hole' test works*, BBC News, Sept. 7, 2015, online at https://www.bbc.co.uk/news/science-environment-34179846 accessed Oct. 10, 2018.
6. Robin McKie, "The future is closer than you though," *The Observer – The New Review*, Sept. 23, 2018, pp. 25-27.
7. Baker, D., "Breaking up is hard to do," *SpaceFlight* magazine, Vol 60, No. 9, Sept. 2018, p. 14.
8. Foust, J., "Advisory group sceptical of NASA lunar exploration plans", *Space News* Nov. 16, 2018, online at https://spacenews.com/advisory-group-skeptical-of-nasa-lunar-exploration-plans/ accessed Jan. 12, 2019.
9. Davenport, C., *The Space Barons*, PublicAffairs (New York), 2018.
10. Solon, O., "Elon Musk: we must colonise Mars to preserve our species in a third world war," *The Guardian,* March 11, 2018, online at https://www.theguardian.com/technology/2018/mar/11/elon-musk-colonise-mars-third-world-war accessed Oct. 10, 2018.
11. Mars – Inside SpaceX", National Geographic documentary movie, 2018.
12. https://www.youtube.com/watch?v=49eIoaY8pVM accessed Oct. 10, 2018.
13. Klotz, I., Falcon Heavy Nails Debut Flight, Aerospace Daily & Defense Report, online at http://aviationweek.com/space/falcon-heavy-nails-debut-test-flight accessed Oct. 10, 2018.
14. Mouriaux, P-F., "Entre Prouesse Technique et Facétie", *Air & Cosmos*, Dec. 21, 2018, pp20-21.
15. NASA's In Situ Resource Utilization (ISRU) website: https://www.nasa.gov/isru accessed Oct. 10, 2018.
16. https://www.spacex.com/mars accessed Oct. 10, 2018.
17. Mouriaux, P. F., *Air & Cosmos*, No. 2582, Feb. 16, 2018, p. 15.
18. Henry C, "Blue Origin enlarges New Glenn's payload fairing, preparing to debut upgraded New Shepard," *Space News* Sept. 12, 2017, online at https://spacenews.com/blue-origin-enlarges-new-glenns-payload-fairing-preparing-to-debut-upgraded-new-shepard/ accessed Oct. 10, 2018.
19. Morring, Jr., F., "Blue Origin's 'New Glenn' Launcher Competes With ULA's Vulcan," *Aviation Week & Space Technology,* Sep. 14, 2016, online at http://aviationweek.com/new-space/blue-origin-s-new-glenn-launcher-competes-ula-s-vulcan accessed Oct. 10, 2018.
20. Zubrin, R., "Lunar Gateway or Moon Direct?", *Space News* April 17, 2019, online at https://spacenews.com/op-ed-lunar-gateway-or-moon-direct/
21. Foust, J., "Blue Origin unveils lunar lander", *Space News* May 9, 2019, online at https://spacenews.com/blue-origin-unveils-lunar-lander/
22. Dunn, M., "Pence calls for landing US astronauts on moon in 5 years", Washington Post, March 26, 2019.
23. Foust, J., "NASA seeks additional $1.6 billion for 2024 Moon plan", *Space News*, May 13, 2019 on line at https://spacenews.com/nasa-seeks-additional-1-6-billion-for-2024-moon-plan/ accessed May 15, 2019.

24. Koren, M., "Could Trump Really Make It to the Moon in 2024?", *The Atlantic*, May 14, 2019, online at https://www.theatlantic.com/science/archive/2019/05/donald-trump-moon-nasa-pence-bridenstine/589453/ accessed May 17, 2019.
25. Chang, K., "For Artemis Mission to Moon, NASA Seeks to Add Billions to Budget", New York Times, May 13, 2019, online at https://www.nytimes.com/2019/05/13/science/trump-nasa-moon-mars.html accessed May 15, 2019.
26. Wikipedia: https://en.m.wikipedia.org/wiki/Orion_(mythology)

11

China, the Communist Challenger

The race between the United States and the Soviet Union to put the first satellite in orbit, then the first man in orbit and then the first manned Moon landing was repeated in a smaller way in eastern Asia in the late 1960s. Japan and China both strove to broadcast their status as modern industrial economies by placing a satellite in orbit. With the help of technology and assistance from the Soviet Union, China succeeded in launching its first satellite on April 24, 1970. Given the name "The East is Red 1"[1] and broadcasting the song of the same name, it is still in orbit today due to the relatively high altitude its launcher achieved, with the high point of its elliptical orbit more than 1,200 miles (1,900 km) above Earth. However its radio stopped transmitting after about three weeks.

Interviewed on state TV in the early years of the 21st century, engineers who developed the satellite confirmed that interference from political commissars (who were attached to state projects in those days) caused delays, as did difficulties in obtaining equipment and material due to disruption caused by the ongoing Cultural Revolution.[2] From an international point of view the launch of this first satellite was something of an anti-climax at the time because China's historical Asian rival, Japan, had launched *its* first satellite two and a half months earlier.

China and Japan had both suffered enormous damage to their industry and infrastructure in the Second World War. By the late 1960s Japan had rebuilt and become an important industrial player in heavy industries such as ship building, and was already establishing itself as a power in consumer electronics such as

---

[1] Literal translation of "Dongfanghong 1."

[2] Interviews witnessed by the author in 2006 on English language state television in China.

portable radios. China aspired to do the same and obtained massive technical and financial support from the Soviet Union. However, its communist dictator, Mao Zedong, imposed policies that were perhaps well intentioned[3] but sometimes ill judged. One of the worst was the ten-year long anti-capitalist, anti-intellectual Cultural Revolution movement that began in 1966, which caused social and economic chaos for the next decade. The anarchy and paranoia of the period meant that even participants in strategic projects such as the first satellite were liable to be arrested and held on spurious grounds.

Having lost out to Japan in 1970 in the 20th century eastern Asia space race, China can perhaps be seen as having overtaken its Asian rival in space when it launched a man into orbit in October 2003. Only the Soviet Union (a role now taken over by Russia) and the United States had previously launched humans into space. Yang Liwei was that first Chinese astronaut, and his framed picture can be seen in the shops and restaurants of Chinatowns around the world, often accompanied by that of Liu Yang, who became the first female Chinese astronaut in June 2012 (see Fig. 11.1).

The Chinese military (the People's Liberation Army) manages China's human spaceflight program, and all of the astronauts to-date have been military pilots. For that and other reasons the United States has found it politically difficult to collaborate with China when it comes to sending humans into space.

NASA Administrator Jim Bridenstine said in 2018 that space is now contested, although he didn't mention China by name. He perhaps had in mind China's 2007 demonstration of its ability to destroy satellites in orbit when it smashed into an old weather satellite of its own with a ground-launched missile, shattering it into thousands of pieces. The quantity of debris in space increased by about 20 percent due to this one action alone, and that debris is still there because it is at an altitude of about 400 miles (600 km) and thus above the region of air drag that would pull it quickly down into Earth's atmosphere. A week after this event China hosted a meeting of a U. N. Committee on the subject of managing the problem of space debris. The deliberate Chinese action to worsen space debris just before the meeting is seen by some commentators as a statement by the People's Liberation Army that its decisions were independent of the agencies that deal with diplomacy.

---

[3] Some of Mao's policies were probably as much to do with strengthening his personal position in power as with improving the Chinese economy.

Fig. 11.1. Yank Liwei *(left)* became China's first astronaut in 2003; Liu Yang became the first Chinese woman in space in 2012. Both are officers in the People's Liberation Army Air Force and former fighter pilots (Illustrations courtesy of Dyor and Tksteven. Used with permission.)

China's 2016 space policy paper [1] states that China "has enhanced standards and regulations" for space debris and goes on to say that "monitoring of and early warning against space debris have been put into regular operation, ensuring the safe operation of spacecraft in orbit." These words suggest that China's military and diplomats have got their acts together and will avoid creating space debris in the future. But the 2007 action remains a wake-up call for other countries, since China is the only country that has ever deliberately generated such a cloud of debris at an altitude where it will survive for centuries.

We saw in previous chapters that both the United States and the Soviet Union toyed with the idea of military-run human spaceflight programs. Both however decided that humans were not terribly useful in space for military purposes. They both undertake extensive military space programs, but have restricted themselves to robotic spacecraft.[4] So why has China's military remained in control of human spaceflight?[5]

One reason is to take advantage of any useful technology that emerges from the human spaceflight program. Much of China's robotic space program is military to

---

[4] The full implication of the Trump Administration decision to create a Space Force as a fourth military branch (alongside army, navy and air force) remains to be seen, but initially it appears to maintain a focus on robotic spacecraft. Some of those spacecraft may have more aggressive missions than is currently the case, such as attacking other spacecraft.

[5] Some commentators dispute the lead role of the Chinese military in the manned space programme, e.g.: [2].

begin with,[6] so it is understandable that when a large new space program comes along the military wish to control that, too. Despite this rather general reason, it is fair to say that Western analysts have not yet been able to figure out the bargaining and competition that has left China's military in control of what seems to be an intrinsically civilian activity. The choice of pilots as astronauts indicates that the air force has considerable influence within the program. This contrasts for example with the lead role of the artillery people in the Cold War Soviet Union space program (see Chapter 8) whose background (firing explosive-packed shells from a cannon) made them uninterested in human spaceflight.

So far, China's manned space activities have been focused on gaining the experience needed to build a laboratory in space. The first precursor laboratory, *Tiangong-1,* reached the end of its five-year useful life in 2016 and fell back to Earth in 2018. Its successor, *Tiangong-2,* has been in orbit since September 2016 and visited a month later by two astronauts who stayed for a month. Since then it has been visited by unmanned spacecraft, which have refueled it robotically – an important skill for any long- term laboratory – and has been in a sort of hibernating state.

The 2016 White Paper states that these activities are to prepare for "exploring and developing cislunar space," where "cislunar" means out to and including the Moon.

A new Tianhe (*Harmony of the Heavens*) space station core module will contain living quarters for China's astronauts and will be launched "in about 2020" – its timing depends on a new rocket, Long March 5B, being available. Later, two science modules will join it, and later still a giant space telescope said to be similar to the American Hubble Space Telescope. The telescope would be in a nearby orbit to the station and would dock for maintenance and repairs.

Also in the early 2020s, China plans to launch a test version of a new-generation human spacecraft. It will be larger than the existing Shenzhou capsule and will be designed for human lunar landings and other potential deep space missions, including to near-Earth asteroids and perhaps even to Mars [4]. The picture one gets is of a slow but steady increase in the scale of human spaceflight with some very ambitious end objectives, with a manned Moon landing as a real possibility.

The Moon is also the focus of the science aspects of China's space program. Whereas other countries had made science an important aspect of their space agenda, that was much less the case for the first 30 years or so of China's space work. However, in the 21st century, China has begun to build ambitious and expensive scientific space probes, and especially those that target the Moon. There is anecdotal evidence that the political dimension of such high profile missions is more important than their scientific potential.[7]

---

[6] See for example [3].

[7] Such as senior politicians turning up for the launch or other media-frequented milestones of such missions but otherwise showing no interest in them.

They placed a robotic rover called *Yutu* (Jade Rabbit) with its *Chang'e 3* carrier ship on the Moon's surface in December 2013, which became the longest functioning spacecraft on the Moon's surface when it finally gave up the ghost fourteen months later in March 2015. In 2018 China has undertaken a daring scheme to land a similar rover, called *Yutu* -2 and its *Chang'e 4* carrier ship, on the hidden face (far side) of the Moon, out of sight of Earth. Its landing site is the Von Karman crater in the heart of the south pole Aitken feature, a large depression that is thought to be the result of an impact early in the Moon's life (Fig. 11.2) and is thus of special scientific interest. Chang'e 4 (with the Yutu-2 rover onboard) was launched on December 8, 2018. It went into orbit around the Moon on December 12, and descended to the surface on January 2, 2019, deploying the Yutu-2 rover 12 hours later. "It's a small step for the rover, but one giant leap for the Chinese nation" quipped the Chief Designer of the overall Lunar Exploration Program, We Weiren, on state TV.

The first step in this pioneering mission had been taken seven months earlier by placing the Queqiao (*Magpie Bridge*) satellite in position 40,000 miles (65,000 km) high above and behind the Moon so that it can see the rover on the hidden side and also see Earth. The satellite can therefore act as a relay point for data and images from the rover to Earth, and radio commands in the other direction. Launched on May 21, 2018, it reached its relay orbit on June 14. China had practiced placing a spacecraft in this unusual orbit seven years earlier, using the *Chang'e 2* probe that had completed its main mission of mapping the Moon in detail.

The 308 lb (139 kg) Yutu-2 rover (and indeed the whole *Chang's 4* mission) is controlled from the Beijing Aerospace Control Center. It goes into a special low power mode during the 14-day lunar night (very cold) and into a temperature control mode for a couple of days in the middle of the day (very hot). Its maximum speed across the ground is 220 yards (200 m) per hour, and it can climb a 20-degree slope and deal with an obstacle up to 8 inches (20 cm) high.

Beyond rovers on the Moon's surface, in the year of *Apollo 11*'s 50th anniversary, 2019, China plans to be the only country bringing actual Moon dust and rocks back to Earth robotically. We saw that the Soviet Union managed this difficult technical feat three times in the 1970s (Chapter 8), but no one has tried to achieve it since then.

China's *Chang'e 5* "lunar scooper" mission will be more complex than those of the Soviets and more like that of the Apollo manned missions. The Soviets landed the whole Luna probe on the Moon's surface and propelled the return capsule direct from there to Earth. China will leave a spacecraft in orbit around the Moon and send a capsule down to the surface, where it will collect samples. The sample container part of the lander will launch itself back to meet up with the orbiting spacecraft, where it will transfer its samples to the special return capsule. The orbiting spacecraft will then return towards Earth carrying the return capsule. The return capsule will separate from the spacecraft and re-enter Earth's atmosphere in a controlled manner while the spacecraft burns up.

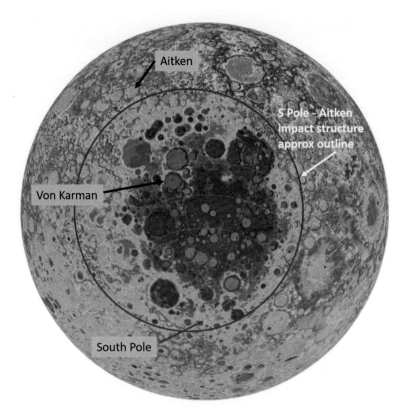

Fig. 11.2. The 1,400-mile-wide south pole-Aitken impact structure is the largest such feature on the Moon and is on the hidden side, never seen directly from Earth. The colors show the depth of the surface below the average surface level. Cooler (bluer) colors represent lower altitudes. China's *Chang'e 4* probe landed in Von Karman crater, a mid-latitude crater with a lava-covered floor that makes it relatively flat and smooth, and thus a safe landing site. There is speculation that the crater was also favored as the landing site because Professor Von Karman, after whom it is named, was the Ph.D. supervisor of the man widely regarded as the founder of China's space program, Qian Xuesen. (Illustration courtesy of NASA/GSFC/Arizona State University/author.)

Successfully returning a capsule from the Moon is much more difficult than returning it from orbiting Earth because of the much higher velocity when re-entering, that is to say hitting the top of Earth's atmosphere – about 24,700 mph (40,000 kph) as compared to 18,000 mph (29,000 kph) [5]. China tested out the technology with a special mission in 2014, called the *Chang'e 5* Flight Test Device (usually shortened to *Chang'e 5* T1). One part of the mission involved flying around the Moon and returning to Earth, at which point a capsule separated from the main probe, re-entered the atmosphere and was successfully recovered on the ground. Meanwhile the main *Chang'e 5* T1 spacecraft returned to the Moon and took up the position behind and high above the Moon, where it could relay data from a probe on the hidden far side. This was a repeat of the exercise already demonstrated by *Chang'e 2* and being used operationally by the 2018 probe Queqiao as part of the *Chang'e 4* far side rover mission. Having performed that exercise successfully *Chang'e 5* T1 then went into orbit around the Moon and practiced rendezvousing with another (virtual) probe, rehearsing for the rendezvous of the *Chang'e 5* lunar sample probe with the mothership that will be orbiting the Moon. Finally the T1 test mission performed detailed photography of potential landing sites for *Chang'e 5* [6].

The thoroughness of the preparation by China for these sophisticated space missions is impressive. From the *Chang'e 1* Moon probe in 2007, China has steadily and gradually increased the capability of its Moon probes, to the point where *Chang'e 4* in 2018/19 is exploring the surface of the Moon's hidden far side – a world first.

A Moon mission of the complexity of the *Chang'e 5* sample return mission is surely a fitting way to celebrate the *Apollo 11* achievements of 50 years earlier, despite the irony that it is being undertaken by one of America's current great rivals.

One possible reason that the sample return mission might miss the *Apollo 11* + 50 deadline in 2019 is that it relies on a new rocket, the Long March 5, which has had a checkered career so far. Its first flight in 2016 was successful, but the second a year or so later failed. There will need to be at least one successful launch of the Long March 5 before China will risk placing the lunar sample return mission on it.

The Long March 5B rocket needed to place the new Tianhe space station in orbit is sufficiently similar to the Long March 5 that it too must await the successful return to flight status of that rocket.

This brings us to the story of China's plans for building very powerful launchers – the single most important issue when it comes to sending humans to the Moon.

The workhorse rockets used by China to launch its satellites into space (see Fig. 11.3) were derived from its long-range missiles, much of the technology for it coming from the Soviet Union. The rockets are usually referred to as the Long March family, and in line with their military heritage they all use storable propellants rather than more powerful fuel combinations that require refrigeration.

However the new generation of Chinese rockets, Long March 5, 6 and 7, have switched to using liquid oxygen with kerosene, and in the case of Long March 5 liquid hydrogen replaces the kerosene in the first and second stages of the rocket. The decision to use fuels that are not much use for missiles illustrates that space has become sufficiently important to justify investing in dedicated engine technology. The distinction between the members of the 5/6/7 family is their power.

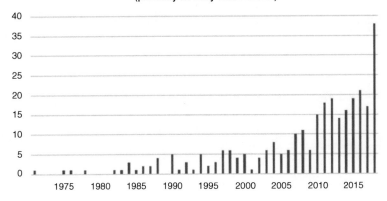

Fig. 11.3. China has been launching 15-20 satellites per year since 2010. The increase over previous years is due mainly to deploying constellations of surveillance and navigation satellites. There have been six missions carrying astronauts, one each in 2003, 2005, 2008, 2012, 2013 and 2016

Confusingly Long March 6 is the baby of the family, capable of placing a 1-ton[8] payload into Earth orbit. (An enhanced version able to loft 4 tons into orbit is planned.) Long March 5 is the heavyweight, capable of placing 25 tons into Earth orbit and of sending more than 8 tons out towards the Moon. In between is Long March 7, able to place 13½ tons in orbit.

As noted above, the heavy-lift Long March 5 is required for the Moon sample return mission, and for the new large space laboratory.

China has built a new launch site as close to the equator as it can in order to increase the performance of the launchers that will be used for Moon missions. Earth spins fastest at the equator, so you get a bit of extra lift if you launch with the spin from there. The new site is on tropical Hainan Island and looks set to rival Cape Canaveral as the top tourism destination for the space enthusiast. Visitors can go inside the site, unlike the situation with the other military-run launch sites on the Chinese mainland (see Fig. 11.4). Couple this with the sandy beaches, warm seas and tropical climate, and the attraction is plain to see.

Early development work is already underway on the super-heavy-lift Long March 9 rocket that will be able to launch about 140 tons into Earth orbit. A prototype of the engine that will power it is being prepared. It will have about four

---

[8] There are at least three different weights that are pronounced "ton." For brevity, I use the word "ton" to signify a weight of 1,000 pounds (about 2,205 pounds) instead of "tonne" or "metric ton." Note that in the United States and Canada, "ton" usually means 2,000 pounds, while in the rest of the world it usually means 2,240 pounds.

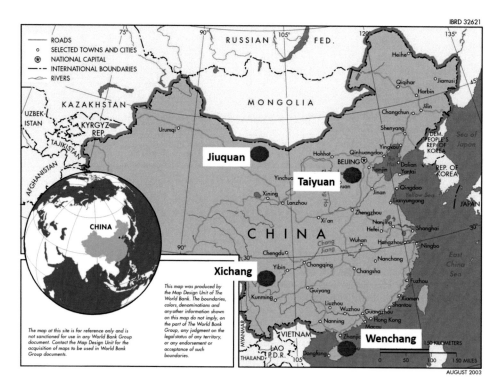

Fig. 11.4. China's four main launch sites. Human missions have been launched from Jiuquan but may switch to the new Wenchang facility on Hainan Island when it is complete and when the new Long March 5 rocket has proved its reliability. Deep space missions, including to the Moon, may also switch there from Xichang

times the thrust of the engine that powers the new Long March 5 heavy-lift rocket. This is consistent with Long March 9 being in the same class as the Saturn V that took Apollo to the Moon. It is tentatively planned to fly for the first time in the 2028 to 2030 timeframe and to be used as part of crewed missions to the Moon and for a robotic Mars sample return.

The Chinese government has not yet approved the full development of the massive Long March 9 and will presumably await the results of the prototyping work and of the current ambitious robotic missions to the Moon. Nevertheless the roadmap leading to a Saturn V class launcher has been publicized, raising expectations at home and abroad.

One feature of China's Moon probes sets it apart from other countries – naming the probes after figures from China's popular culture. While U. S. and Soviet probes had functional names such as Surveyor, Ranger, Lunar Orbiter, Lunokhod and Luna, the Chinese probes have names that resonate with the public. The series of craft sent to the Moon since 2007 have been called Chang'e, which is the name of the Chinese Moon goddess. *Chang'e 1* (launched 2007) and *Chang'e 2*[9] (2010) orbited the Moon, taking detailed photographs of potential landing sites for *Chang'e 3*, which landed on the Moon in 2013 carrying a rover. The *Chang'e 3* rover was given the name Yutu, or jade rabbit, who in legend accompanied the goddess Chang'e to the Moon (see Fig. 11.5).

Fig. 11.5. Model of the 310 lb (140 kg) Yutu rover that landed with *Chang'e 3* on the Moon on December 14, 2013. It took photos and made measurements of the surface, but it is not clear how far it actually moved. The Yutu-2 rover on *Chang'e 4* is broadly similar but lacks a robotic arm. (Illustration courtesy of Brian Harvey. Used with permission.)

The relay satellite for the *Chang'e 4* rover on the far side of the Moon is named Magpie Bridge (Queqiao). The Chinasage website [7] relates the legend that explains why this name is so appropriate. On the 7th day of the 7th lunar month[10] a bridge of magpies allows the two lovers Zhinu and Niulang to be briefly reunited among the stars. The legend allows the bridge to exist for just one day each year,

---

[9] After eight months, *Chang'e 2* fired its engines to leave its orbit around the Moon and head into deep space, first to a point where Sun and Earth gravity neutralize each other (the 2nd Lagrangian Point). Then ten months later it proceeded outwards to fly by the asteroid Toutatis.

[10] August 7 in 2019, August 25 in 2020 for example.

but hopefully the 21st century Queqiao will provide a communications bridge every day. The *Chang'e 4* landing site has been named Milky Way Base, and nearby craters have been given names from Chinese folklore one being that of the separated lover Zhinu.

The *Chang'e 4* mission is using social media to build public interest in the mission. A microblogging site on Weibo (the nearest Chinese equivalent to Twitter, which is blocked from public use) includes regular posts from Yutu-2 such as "It is getting hot here," "I have to take a siesta for a while," "My masters have given me insulating components," "I feel proud that even the small Chinese flag on my chest can withstand high temperatures," "The heat control abilities of my fourth sister (its name for the *Chang'e 4* carrier ship) are stronger than mine," and "Master, remember to wake me up early when the work starts again." During the "noon siesta," Yutu-2 will write a "diary," send monitoring footage and provide readers with stories about the Moon [8]. We may expect future lunar probes to bear other names from Chinese legends such as Wu Gang, the woodcutter who is doomed to live forever on the Moon trying to cut down the self-healing laurel tree.

The Moon goddess is a much loved figure in Chinese culture and is the focus of the annual mid-Autumn Moon or Harvest Festival – second in importance only to New Year in the Chinese calendar. Mooncakes are traditionally eaten during the festival, and have become known in the West as a sweet delicacy. Although the harvest is no longer so central to Chinese life, the Moon Festival is still widely celebrated as a time to reunite with friends and relatives, recognizing the Moon as symbolizing harmony. Public enthusiasm for a government policy (especially an expensive one) is always a sensible political tactic, and this naming policy is a cost effective way to help achieve that. The willingness of the Chinese authorities to tap into traditional culture by naming lunar probes after popular legendary characters suggests a strong commitment to pursue exploration of the Moon.[11]

Chang'e and Yutu had a walk-on non-speaking role in the *Apollo 11* mission. On July 20, 1969, a minute or two before Armstrong and Aldrin entered the Lunar Module to begin their descent to the lunar surface, Mission Control in Houston asked them to keep an eye out for a legendary Chinese lady and her pet rabbit: "Watch for a lovely girl with a big rabbit. An ancient legend says a beautiful Chinese girl called Chang-o has been living there for 4,000 years. You might also look for her companion, a large Chinese rabbit, who is easy to spot since he is always standing on his hind feet in the shade of a cinnamon tree." Michael Collins [9] was quick to respond with "We'll keep a close eye out for the bunny girl."

The Moon program is yet another way that China is using to establish itself as a regional superpower. Its aggressive attempts (as seen by some of its neighbors)

---

[11] The policy of using mythical names for space systems was apparently initiated by Deng Xiaoping, the wily Chinese leader who set China on its present course of competing in the global market economy in the 1980s. Previously, names from the communist party's history were used such as Long March.

to redraw maritime borders in the South China Sea and in the East China Sea are evidence of the country's increased power. No longer abiding by Deng Xiaoping's advice to "Hide your strength and bide your time," the catch phrase under current leader Xi Jinping has been "the great resurgence of the Chinese nation," admittedly downplayed somewhat in the wake of U. S. import tariffs [10, 11]. Space spectaculars such as the Moon probes and the astronaut missions are a soft way to demonstrate your superiority over neighboring countries without threatening their day to day interests. Undertaking space missions that India, for example, is not able to copy sends an unmistakable message to the inhabitants and rulers of all Asian countries that China is technically and financially the leader.

President Xi Jinping referred in April 2018 to the great achievement of launching a satellite in 1970: "By tightening our belts and gritting our teeth, we built [the atomic and hydrogen] bombs and one satellite." President Xi is perhaps the first Chinese leader in history with a global vision, and he inevitably looks to the United States as the yardstick against which to measure a country's performance in the superpower stakes. According to China expert Professor David Shambaugh, he recognizes that the Soviet Union's fundamental mistake in its attempt to maintain that status was overreliance on military power. Xi appears to understand that a strong economy, technology and soft-power influence are needed for long-term sustainability as a global leader. Wang Wen from Renmin University in Beijing says that "President Xi has repeatedly emphasized that China must participate more actively in setting the rules in new areas including the internet, the deep sea, the polar regions and outer space [12]."

Xi's strong support for space as an area on which to focus is evident by his sponsorship of space managers in the political arena. Four provincial governors (including Guangdong, the province with the biggest economy) were formerly in the space industry. In particular, in 2017 former China Academy of Space Technology president Yuan Jiajun, who was also chief commander of China's Shenzhou manned space program, was appointed acting governor of Zhejiang, a hub of private business and the power base of President Xi – giving a new meaning to the term "high flyer [13]."

China's modest human spaceflight activities and its ambitious Moon probes have given it a distinct and sophisticated space presence that ensures it will help to shape the rules for the future use of space. Furthermore, that distinct and impressive China-in-space image leaves it well ahead of Asian rivals in that respect. India's belated entry into human spaceflight (see Chapter 12) is perhaps a recognition of how distinct China's space activities have become, and how they have established China as the Asian leader in space. As the 2020s progress, China will be able to maintain that position if it continues to work towards a human Moon landing coupled with complex robotic Moon-exploring machines, since only the United States seems able to undertake a similar mix of space missions.

Finally we should not ignore the possibility of China and Russia collaborating on a project to send humans to the Moon. Western politicians and commentators were notoriously inaccurate in their reading of Soviet and Chinese policies and motivations during the Cold War. The Chinese leader Mao Zedong played the West off against the Soviet Union in the 1950s and 1960s in order to get military technology from the latter, including nuclear weapons. Western misunderstanding of the two big communist powers continued throughout the Vietnam War into the 1970s, with the United States wrongly assuming that the Soviets and Chinese could or even wanted to control the Vietnamese communists.[12] How sure are we that we understand Beijing-Moscow thinking in today's world?

Russia and China have had serious disagreements over their 2,600 mile (4,200 km) long common border, sometimes escalating into shooting conflicts. However, in recent years, the two countries have increased trade, and even undertaken joint military exercises (admittedly the Chinese involvement was fairly nominal) during which Russian President Vladimir Putin and Chinese President Xi Jinping shared pancakes and vodka in Vladivostok in a public display of their rapport. In fact, Putin and Xi have met more than 25 times, far more frequently than either has with any other head of state. Some of the reasons for this improved relationship are not hard to understand. Russia is suffering from low prices for its oil exports and from Western trade embargoes, so increased energy sales to China make sense, while China is eager to buy cheap oil and natural gas. Russia is also keen to demonstrate that it is not totally dependent on the West for trade.

There are other more worrying reasons for this new Moscow-Beijing coziness. Both countries are active in hacking Western computer systems and in suppressing human rights at home, reflecting their authoritarian politics. There are of course differences, with Russia currently more aggressively using military power to change borders and achieve political ends. Beijing benefits from this Russian assertiveness which exposes the cracks in Washington's alliances and distracts the United States from China's rise. One Western commentator noted that "the [two] countries' strategies have become mutually reinforcing in powerful, if perhaps unintended, ways [16, 17]."

Is it therefore out of the question that China and Russia would pool their resources in seeking to send humans to the Moon? Neither has to be identified as the leader in the team. They can use the same formula as the United States and Russia use in the ISS, with both recognized as equal partners. There are no concrete signs of this collaboration yet, but if Presidents Putin and Xi Jinping decide it's a good idea, then it is likely to happen.

---

[12] See for example [14, 15].

References

1. State Council of People's Republic of China, *White Paper on China's Space Activities in 2016*, December 28, 2016, online at http://english.gov.cn/archive/white_paper/2016/12/28/content_281475527159496.htm accessed Oct. 11, 2018.
2. Harvey, B. *China in* Space, Springer Praxis (Chichester, UK), 2013, pp. 354-355.
3. "China Aims for the High Ground" by Norris, *Aerospace International*, October 2012, pp. 22-25.
4. Jones, A., "Chinese space program insights emerge from National People's Congress," *Space News*, April 2, 2018, online at (accessed October 8, 2018). http://spacenews.com/chinese-space-program-insights-emerge-from-national-peoples-congress/.
5. Mu Xuequan (Ed.), *China schedules Chang'e-5 lunar probe launch*, Xinhua, Jan. 22, 2017 online at http://news.xinhuanet.com/english/2017-01/22/c_136004958.htm (accessed Oct. 8, 2018).
6. Siddiqi. A., *Beyond Earth: A Chronicle of Deep Space Exploration 1958-2016*, NASA SP-2018-4041, Sep. 2018, pp. 272-273, 291-294, 297-298.
7. http://www.chinasage.info/symbols/birds.htm#XLXLSymMagpie (accessed Oct. 8, 2018).
8. Pinkstone, J., You, T., "It's getting hot here", Mail Online, Jan. 8, 2019, online at https://www.dailymail.co.uk/sciencetech/article-6568513/Chinese-moon-rover-Jade-Rabbit-2-sends-cute-social-media-messages-taking-nap.html?ito=1490 accessed Jan. 13, 2019.
9. Woods, D., MacTaggart, K., O'Brien, F., "Apollo Flight Journal," March 2016, Day 5, *Time* 095:17:28 - 095:18:15 available at https://history.nasa.gov/afj/ap11fj/index.html (accessed Oct. 8, 2018). An earlier version of the transcript attributed the "bunny lady" response to Aldrin.
10. French, H. W., "China's Dangerous Game," *The Atlantic,* November 2014 available at http://www.theatlantic.com/magazine/archive/2014/11/chinas-dangerous-game/380789/?single_page=true accessed Oct. 8, 2018.
11. Crawford, A., "Xi's Summer of Discontent," *Bloomberg Businessweek*, Aug. 13, 2018, pp. 39-41 available at https://www.bloomberg.com/news/articles/2018-08-07/is-xi-jinping-s-bold-china-power-grab-starting-to-backfire accessed Oct. 8, 2018.
12. Champion, M., "China's Pole-to-Pole Ambitions" *Bloomberg Businessweek,* September 3 2018, pp34-37.
13. Jun Mai, "Why China's aerospace experts have become Xi Jinping's new political elite", *South China Morning Post*, May 4, 2017, online at https://www.scmp.com/news/china/policies-politics/article/2092940/how-leaders-chinas-space-programme-entered-political (accessed Dec. 18, 2018).
14. Norris, P., *Spies in the Sky*, Springer Praxis (Chichester UK), 2007, pp. 163-167.
15. Hastings, M., *Vietnam, an Epic Tragedy 1945-1975*, William Collins (London), 2018, p. 477.
16. Kendall-Taylor, A., Shullman, D., "How Russia and China Undermine Democracy," *Foreign Affairs,* Oct. 2, 2018, online at https://www.foreignaffairs.com/articles/china/2018-10-02/how-russia-and-china-undermine-democracy accessed Oct. 12, 2018.
17. Kendall-Taylor, A., Shullman, D., "A Russian-Chinese Partnership Is a Threat to U.S. Interests", *Foreign Affairs*, May 14, 2019, online at https://www.foreignaffairs.com/articles/china/2019-05-14/russian-chinese-partnership-threat-us-interests accessed May 17, 2019.

12

Russia and the Rest

Russia

The Soviet Union was once the most powerful nation in space, but taking on its role at the end of the Cold War, Russia has fallen behind the United States in space affairs. It has been almost thirty years for example since the last even partially successful Russian planetary mission. However, under President Vladimir Putin Russia has been sending messages that it is ready to rejoin the top club. But words are cheap –while going to the Moon is not!

The legacy of the great Soviet space engineers of the 1950s and 1960s still lives on in Russia. Since the United States mothballed the space shuttle in 2011, all crew ferried to the ISS travel in the Soyuz capsule atop the Soyuz rocket, both designed by Korolev's team at the start of the Space Age. And the Proton rocket that carries cargo to the station was designed by Chelmoei's team also in the 1960s.

Some other space legacies of the Soviet Union are less convenient. Russia, alone among the 15 republics that made up the Union of Soviet Socialist Republics (USSR),[1] has nuclear weapons and the missiles that carry them. When the union was dissolved in 1991 the other countries agreed to transfer their nuclear weapons to Russia. The same is *not* true for Soviet space resources. Russia was by far the largest of the 15, and so the bulk of the factories and laboratories that created the Soviet Union's rockets and satellites are in Russia. But some that aren't have proved to be a headache.

The most obvious space activity taking place outside Russia is the Baikonur launch site, which is in Kazakhstan – the equivalent of Cape Canaveral in the United States. All launches of cosmonauts and foreign space travelers take place

---

[1] Armenia, Azerbaijan, Belarus, Estonia, Georgia, Kazakhstan, Kyrgyzstan, Latvia, Lithuania, Moldova, Russia, Tajikistan, Turkmenistan, Ukraine and Uzbekistan.

from here. And about half of all other launches also use this site. The other main launch site is the Plesetsk Cosmodrome about 500 miles (800 km) north of Moscow, which is convenient for launching satellites that go over the poles (a common orbit for surveillance satellites). Russia and Kazakhstan have had a contractual relationship since 1991, with the Kazakhs paid for the use of their facilities.

Fig. 12.1. Russia's three main space launch sites (Cosmodromes): Baikonur, Plesetsk and Vostochny. The vast distance between the new Vostochny Cosmodrome and the space manufacturing facilities in the west complicates its use (international borders shown as in 1994)

Not wishing to be dependent on Kazakhstan forever, in 2007 Russia began developing a new launch site in eastern Russia from which launches to the east out over the Pacific Ocean would be possible. Launching eastward is desirable in order to get the benefit of Earth's spin. The Vostochny site is six time zones away from Moscow, and its development has been beset by delays and corruption scandals (see Fig. 12.1).[2] President Putin stepped in to take personal responsibility for it in 2015, which has confirmed the government's commitment to complete the

---

[2] In March 2018 the boss of the construction company responsible for building the site was sentenced to 12 years in prison.

facility. The first launch took place in 2016 (see Fig. 12.2), with further launches in November 2017 and February and December 2018. Until Vostochny can support more frequent launches, Baikonur in Kazakhstan will remain one of Russia's main launch sites.

Fig. 12.2. The first launch from the new Vostochny Cosmodrome in 2016, a Soyuz 2.1a rocket. On the white building in the foreground is a picture of Yuri Gagarin with the slogan (in Russian) "Raise your head." Roughly one launch per year is the rather sorry average rate at the moment at Vosotchny – nowhere near enough to justify the expense of its development (Illustration from www.kremlin.ru.)

Investment in the Vostochny launch site can be seen as part of President Putin's "pivot to Asia" announced in 2014. The stated aim is to focus Russia's exports on Asian giants such as China, India and Japan, and to reduce the country's dependence on exporting to Europe. It is also supposed to increase economic activity in Russia's eastern regions, which generally are much less developed than regions further west. After several years of much talk and little obvious results, skeptics suggest that the real aim of the Asian pivot is to bolster Russia's standing in its confrontation with the West.[3] Increased sales of military equipment to China and India, plus softening of its hard line in negotiations with Japan about the disputed ownership of the four southernmost Kurile Islands are intended to signal to the

---

[3] See for example [1].

United States that Russia can form alliances with these countries, to the detriment of the United States. In this reading of events, Russia's Asian pivot is a negotiating tactic to encourage the United States and the West to be less hostile to Russia – which seems very possible given the pro-Russian statements of President Trump.

Further investment in the Vostochny Cosmodrome will be easier for President Putin to support if it helps increase Russia's image as a superpower and strengthens its credentials as a major actor in Asia. Were Russia and the West to become friendly again, then Russia would no longer need an Asian pivot, and Vostochny's future might not be so bright. What the effect would be of collaboration between Russia and China in a Moon adventure (see Chapter 11) is even more difficult to predict.

More subtle loose ends of the breakup of the Soviet Union are the facilities in Ukraine. Only two of the 15 countries in the Soviet Union had populations that made up more than 6 percent of that of the union as a whole: Russia with 51% and Ukraine with 18%.[4] It is no surprise therefore that these two countries contained the vast majority of the industrial strength of the union. Ukraine's industry contributed to many parts of the Soviet space program – both the satellites that went into space and the rockets that carried them there. For example the *Tselina-2* satellites that gathered electronic intelligence for the Russian military were built in the Ukraine and continued to be bought by Russia until locally sourced replacements were available. The first launch of the *Tselina-2* replacement, Lotos, was in 2009 [2].

The Ukrainian Zenit-2 rocket has proved more difficult to replace with a Russian alternative. In 1992 Russia decided to develop the Angara series of rockets, which would include a Zenit-2 replacement. Beset with delays and funding difficulties, two versions of the rocket were launched in 2014 (22 years after it was conceived), but none since. Its future remains in doubt, a victim of the many changes in direction of Russia's space program, one of which we will now examine because it concerns missions to the Moon and appears to signal the end of the Angara.

Before looking at Russia's Moon plans, there is another Russia-Ukraine "loose end" resulting from the breakup of the Soviet Union that relates to this story. The two most influential rocket designers of the early Soviet space program, Sergei Korolev and Valentin Glushko (see Chapter 8), were both born in Ukraine. Russia has inherited the bulk of the Soviet space activity and lays claim to the related heritage, including these two historical figures. Ukraine considers both men to be Ukrainian. When Ukraine was part of the Soviet Union that was not an issue, but now that Russia and Ukraine are at daggers' drawn, that historical legacy is another flashpoint in their relationship.

In the background to any discussion about future human spaceflight is the issue of what to do with the ISS after 2025 – a question faced by the United States, too,

---

[4] 1989 population totals: Russia 147 million, Ukraine 52 million, Soviet Union total 287 million.

as discussed in Chapter 10. Broadly speaking Russia has explored three very different schemes:

1. Build an all-Russian space station.
2. Join the United States in setting up the Lunar Gateway.
3. Aim for a go-it-alone lunar landing program.

In May 2017 President Vladimir Putin instructed his space agency, Roscosmos, to give priority to developing a super-heavy lift rocket with a view to a manned Moon landing – essentially choosing the last of these three schemes. The first test launch of the new rocket was set for 2028.

The decision includes the choice of engines to use – and the Angara technology was not selected. Instead the Russian engine used in the first stage of the Ukrainian Zenit-2 rocket was preferred. In case you have lost track: a Russian engine was used in the Ukrainian Zenit-2 rocket when they were both in the Soviet Union. That was fine then, but now that they are in a relationship that is perilously close to war, it's a mess. The Russian plan is to first develop a Zenit-2 class rocket called Soyuz-5. Then they will move on to build the super-heavy lift rocket, for which the first stage will be a cluster of five engines from Soyuz-5. The second stage of the rocket would be a single engine again of the same type. The third stage would be comprised of a new sophisticated engine.

Overall, by reusing existing technology, the amount of development needed to create a rocket that can launch 100 tons[5] or more into Earth orbit is not too daunting – at least on paper.

Putin has held four dedicated meetings with the country's top space managers and paid two visits to the new Vostochny Cosmodrome in the far east in the last few years – a level of presidential interest in space far above that in the United States (see Fig. 12.3). He seems committed to making the country a space power again, but so far has not followed through with the increased funding consistent with that goal. And given Russia's status as one of the world's major suppliers of oil, several years of low oil prices have reduced central government revenues, making any budget increases difficult. Indeed a drop in Putin's popularity in mid-2018 has been the result of his plans to raise the age at which the state pension will be paid, reflecting the government's budget difficulties.

Skeptics, and there are many, point out that this is the fifth attempt to define a strategy for Russia's space program since 2012: a plan for space activities in the period 2013–2020 (approved in December 2012), a national space strategy until 2030 (signed by Putin in April 2013), the Federal Space Program 2016–2025 (ratified in March 2016), and a Roscosmos strategy until 2030 (the policy that includes

---

[5] There are at least three different weights that are pronounced "ton." For brevity, I use the word "ton" to signify a weight of 1,000 pounds (about 2,205 pounds) instead of "tonne" or "metric ton." Note that in the United States and Canada, "ton" usually means 2,000 pounds, while in the rest of the world it usually means 2,240 pounds.

Fig. 12.3. Russian President Vladimir Putin visiting the Vostochny Cosmodrome in 2015. Corruption scandals involving officials and the construction companies have delayed the completion of the site (Illustration from www.kremlin.ru.)

the super-heavy lift rocket). As one seasoned analyst of Russia's space program put it: "Few outside the Russian space community seem to understand how these documents are interrelated, how exactly they come about, and how binding they are [3]."

The funding of the program is the key issue. Initial funding for the first step in the development has been provided. That first step is to develop the Soyuz-5 rocket (confusingly also referred to sometimes as Sunkar[6] and Feniks) to replace the Zenit-2 rocket. The main engine of the Zenit-2 was already built in Russia, so by using that engine the development is considered fairly low risk, with a first flight

---

[6] Sunkar is a Kazakh word that means "falcon," which Russian officials say relates to the importance of falconry in Kazakh culture, not to the SpaceX Falcon rockets.

scheduled for 2022. Changes will be required to the engine so that it incorporates only Russian components, partly dictated by U. S.-imposed sanctions. Mind you that Russian engine has been out of production for several years, so there could be unexpected delays getting it up and running again. It is in fact the world's most powerful rocket engine, and scaled down variants of it have been sold to the United States to form part of America's Atlas and Antares launchers. A contract valued at about $800 million (52.7 billion rubles) has been placed with RKTs Progress, the company tasked with developing the Soyuz-5. Continued funding seems to be tied to Soyuz-5 being able to win commercial launch contracts, and that will be increasingly difficult as SpaceX drives down the price of launchers.

The super-heavy lift rocket is being referred to by its Russian acronym STK. Funding to create its massive cluster of engines has not yet been guaranteed. Given that the first and second stages will use the engine being developed for the Soyuz-5, the development of the third stage engine could be the pacesetter for the whole rocket. It is to be fueled by liquid hydrogen and liquid oxygen, which we have seen before is an extremely powerful fuel combination but requires ultra-sophisticated refrigeration equipment and plumbing. The last Russian rocket engines using that fuel were flown in 1987-88. The same company that built those engines, KBKhA, will also build the STK third-stage engines, but much of the expertise has been lost in that thirty-year gap.

The launch sites from which the Soyuz-5 and later the massive STK will be launched also require major investment. Soyuz-5 will be launched from the long standing Baikonur facility in Kazakhstan, on the assumption that the Kazakhs will contribute to the refurbishment and upgrading of the existing facilities. STK will be launched from the new Vostochny Cosmodrome in the far east of Russia, whose development as we have seen has been marred by corruption scandals. A huge range of new facilities will be needed there for the STK, reminiscent of that required at Cape Canaveral for the Saturn V in the 1960s. This will include an almost 400-foot-high assembly building, a rocket storage building, two spacecraft integration buildings, a new launch pad that can accommodate the STK and the Soyuz-5 as well as handle test firings of the STK stages, and a mobile service tower [3]. Having these facilities in place by the mid-2020s is very ambitious (some would say unrealistic), but that is the schedule if the first STK launch in 2028 is to be met.

There are challenges, too, concerning the spacecraft to be carried by the STK to the Moon. The new Federation[7] spacecraft will carry up to four cosmonauts to the Moon (up to six into Earth orbit) and is intended to replace the Soyuz spacecraft that has been ferrying humans to and from space for 50 years. Originally contracted to industry in 2009, it has been subject to changes in its function (for example initially only to orbit Earth, now required to also orbit the Moon) that have delayed its construction. Federation is designed to land more accurately than

---

[7] *Federatsiya* in Russian.

Soyuz and thus avoid the need to return to the vast open landscape of Kazakhstan, as it will land on Russian territory. As of early 2019, its first test flight is scheduled for 2022 from Vostochny without any crew, leading to manned flights in 2024 and an unmanned flight to the Moon in about 2025. But even if it meets that schedule, will its Soyuz-5 rocket and the Vostochny Cosmodrome from where it will be launched be ready for those dates? The odds of that happening are not good.

A generation has been lost for the space industry, during its struggle to survive. The staff employed in the space sector are now either over 60 or under 30, with almost no intermediate age group. Moreover, Russia has largely abandoned the Soviet practice of inspection of space products by independent inspector teams composed of military officers with engineering training. This has led to fabrication and procedural errors beginning to appear. A recent example was the small hole drilled, apparently by accident, in the hull of the Soyuz spacecraft detected while it was attached to the International Space Station. In the view of one analyst, under Putin the surest way to get government funding has been to propose bold and ambitious space missions that did what no other country had done before, but the loss of expertise has led to a Catch 22 situation where such missions are now beyond their engineering capability to accomplish [11].

In summary, Russia has stated that it will develop a very powerful rocket that will allow it to send men to the Moon by about 2030. However its recent track record for delivering new space programs is poor, and its economy is suffering not least from low oil prices. One respected analyst, Anatoly Zak, says that "Putin's bold and costly move comes at a low point in the Russian space program, where the unending string of failures, years-long project delays, and corruption scandals show signs of continuous dysfunction within the Russian space industry." He goes on to say that "there are signs that the crisis is actually deepening, especially as Russia faces an uphill battle on the international launch market. Building the super-heavy rocket is a controversial way to resolve the problem [4]."

The attempt to beat Apollo to the Moon in the 1960s was plagued by frequent changes of direction and drip feed funding. It looks like not that much has changed in the 50 years since, so why should the outcome be any different, with Russia likely to end up again in the DNF[8] category. Unless it teams up with China – see Chapter 11.

Other Countries

The other countries spending serious money on human spaceflight are the junior partners in the ISS: several European nations,[9] Japan and Canada, plus India. Europe and Japan can launch robotic spacecraft, but unlike the senior partners

---

[8] Did not finish.

[9] Mainly through ESA, an international agency not legally tied to the European Union but representing pretty much the same countries.

(the United States and Russia) have no rockets certified for human missions. Let's do a quick sanity check on these countries to see if they will have an impact on the next humans to walk on the Moon's surface.

Europe

Germany triggered the Space Age with the development of the V-2 rocket during World War II. The V2 reached an altitude of about 125 miles (200 km) in test flights,[10] which is well beyond the 62-mile (100-km) boundary where space begins, the first time a manmade object achieved that.[11] But after the war, the technology and many of the engineers emigrated (not always voluntarily) to the United States and the Soviet Union, not least Wernher von Braun, who went on to lead the development of the Saturn V rocket in the United States that was critical to the Apollo program.

Britain and France developed medium-range missiles to carry their nuclear weapons but never deployed them in the vast numbers that the United States and Soviet Union superpowers did. European countries were content to buy American rockets to launch their satellites into space until the end of the 1960s, when the United States began to refuse to launch some satellites on the grounds of international law (U. S. version) or to favor competing U. S. products (European version). Europe therefore developed its own space launcher, the Ariane series, thus avoiding having to rely on non-European rockets.

The military focus of early European rocket development meant that funding to send humans into space was low on the list of the budget priorities of governments. Nevertheless, after the drama and excitement of the Apollo Moon landings, Europe decided to play a small role in America's next human spaceflight adventure, the space shuttle. The European contribution to the shuttle was a laboratory called Spacelab that was carried in the shuttle cargo bay and enabled the astronauts to undertake scientific research during the typically week-long shuttle mission. This was a toe-in-the-water for Europe into the expensive business of human spaceflight.

Europe then decided to get its ankles wet in the next American manned spaceflight exploit, the ISS, and has contributed about a tenth of the station funding mainly in the form of station modules that the astronauts can live and work in. About half of the station's habitable space has in fact been supplied by Europe. Europe also provided some cargo supply flights to the station using its Ariane rocket.

While signing up to participate in the ISS Europe considered going all the way into human spaceflight but then backed off. The idea was to certify the Ariane

---

[10] Its peak altitude when fired operationally was about 80 kilometers.

[11] The boundary of space is a subject of debate, with some experts proposing a value of 80 km (50 miles), see [5].

rocket so that it could carry humans and to build a winged spacecraft called Hermes in which the astronauts would be carried into and from space. The price tag for this ambitious development proved too steep for European governments, and they settled for a junior role in the U. S.-led station. There have been some recent wistful statements from industry leaders about Europe making its newest rocket, Ariane 6, safe enough for launching humans. but the logic of developing Ariane 6 was to *reduce* the cost of launches in order to compete with America's SpaceX, whereas human-rating a launcher would make it more expensive.

On the robotic front, Europe has shown little interest in the Moon, instead focusing its attention on comets, Mars and general astronomy. European scientists have tended to favor these more distant targets, and it is difficult for governments to fund robotic exploration of the Moon without its scientists' backing.

As the ISS nears the end of its life, Europe has already signed up to provide a major part of NASA's Orion spacecraft that will be carried by the Space Launch System (SLS) to the Lunar Gateway. The Service Module to be supplied by Europe will provide Orion's propulsion, electrical power, water, oxygen, air conditioning, etc. Each Service Module incorporates an engine supplied by the United States, but still costs Europe about $250 million (see Fig. 12.4).[12]

In late 2019 European governments will decide whether to sign up for a role in the Lunar Gateway itself. The objective would be to have an occasional European astronaut at the gateway, to manage robotic lunar rovers and probes from there, to supply more Service Modules for the Orion spacecraft and to test technologies for human exploration on the Moon's surface – this last is being studied jointly with Japan and Canada. The scenario calls for Europe to divert its current station funding of about $450 million a year to the Gateway program. This means that funding of the station would have to stop in 2025 and that extra funding would be needed between now and then to get started on Gateway activities. There had been pressure in Europe to keep funding the station, but Russia's central role in it has made the station much less politically attractive because of Putin's militaristic foreign policy.

Europe is unlikely to sign up in 2019 to actual NASA-led manned Moon landings but will probably agree to consider missions that are tentatively scheduled for beyond 2028, with a decision by Europe to participate or not by about 2023.

European astronauts who have been to the ISS attract great public interest in their home countries, including at the top political level. A recent example was Frenchman Thomas Pesquet who, following his return from the station in June 2017, soon became an informal adviser to French president Emmanuel Macron. So, provided the NASA Lunar Gateway plan survives intact until late 2019 (not a

---

[12] The first cost about $450 million (despite lacking some functions such as the ability to provide a breathable atmosphere for the crew cabin), the second about $230 million.

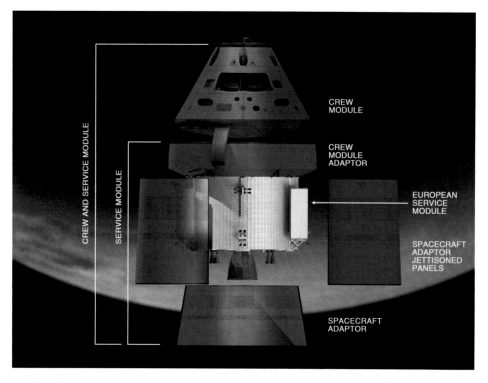

Fig. 12.4. Europe is supplying the Service Module for the first two of NASA's Orion spacecraft that will take astronauts to and from the Lunar Gateway. The Service Module provides power, propulsion, air, water, cooling, heating and other utility functions. Europe's role is a payment in kind for part of its 8% contribution to the costs of the ISS. Funding for further Service Modules would be part of a European commitment to the Lunar Gateway. (Illustration courtesy of NASA.)

certainty given the SpaceX and Blue Origin initiatives), and provided that an end to the ISS can be agreed, it is likely that Europe will sign up for a Gateway role.

Japan

Without nuclear weapons that require long- range missiles, and with its military constrained to a defense-only role by the treaty that ended World War II, Japan has had to fund the development of its space rockets from strictly civilian budgets. The resulting rockets have been relatively expensive and almost exclusively used to launch Japanese government satellites.

Japan's role in the ISS has been similar to that of Europe – building one of the laboratory modules of the station and launching cargo to the station every 18 months or so. Funding has been about 20 percent less than Europe's, running at about $360 million a year. Its astronauts who return from the station are inevitably media stars generating huge public interest.

Political tensions in eastern Asia have if anything served to strengthen Japan's ties to the United States, especially given the nationalistic stance of Prime Minister Shinzō Abe. His proposed removal of the constitutional ban on Japan's defense forces being used outside the country, plus his perceived reluctance to apologize for Japan's actions during World War II, make him unpopular in China, Korea and other countries occupied by Japan during the war. Japan's dispute with China over some barren rocks in the East China Sea aligns with America's attempts to restrict Chinese expansion in that area.

So it may well be politically desirable for Japan to remain part of America's flagship manned spaceflight adventures, negotiating a junior role in the Lunar Gateway. The tricky issue of closing down the ISS and ramping up the funding for the Gateway until the station closes applies just as much to Japan as to Europe (and the United States).

Canada

Canada has carved out a niche role for itself in first the space shuttle and now the ISS, providing a sophisticated robotic manipulator for both. The Canadian robotic arm (dubbed "Canadarm") is an essential piece of equipment on the station, used whenever work on the outside of the station needs some heavy lifting. That includes moving an astronaut and his or her tools around as part of a spacewalk. The price tag for Canada's involvement in the station currently runs at about $45 million a year.

Canadian astronauts have made a name for themselves. Julie Payette spent a total of three and a half weeks in space on two space shuttle visits to the ISS in 1999 and 2009. On both occasions she took charge of the Canadarm operations. She is now Governor General of Canada, acting as the official representative of Queen Elizabeth 2 in the country. Chris Hadfield's five-month stay on the ISS in 2012-2013 gave a boost to his potential career as a country and western singer thanks to his live video broadcasts from space, and also led to a reality TV show series.

It seems likely that a role as provider of robotic equipment for the Gateway would be of interest to Canada, subject as for other countries to the funding arrangements being acceptable.

As noted above, Canada is already working with Europe and Japan to design a series of robotic probes to the Moon's surface, with Canada especially involved in a possible robotic rover. And in Chapter 10 we saw that Canada was working with Europe on remote control of robots on the Moon, using parts of Quebec province as a proxy for the lunar landscape. These initiatives give Canada and its partners a route to exploring the Moon that doesn't require actual boots on the ground there. It can be a steppingstone towards sending humans there or can be a fully-fledged alternative.

India

In August 2018 Indian Prime Minister Narendra Modi chose a highly symbolic date and venue to make two major policy announcements. The date was Independence Day, August 15, which celebrates the formation of the country when it gained freedom from the British Empire. The venue was the Red Fort in the nation's capital, New Delhi, which was the main residence of the last indigenous Indian empire – that of the Mughal dynasty.

The first policy was the start of a national medical insurance scheme, dubbed "Modicare," which plays well to the mainly poor Hindu segment of the population that support Modi. The policy had already been announced, and the new element was a date that it would begin: September 25, 2018.

The second policy was a commitment to launch an Indian citizen into space in time for the 75th anniversary of independence from Britain in 2022 (a mere four years later). The plan is for three Indian astronauts to spend a week or so in orbit around Earth.

By announcing the astronaut program on such a high profile occasion, Modi put the prestige of his government behind it being successful. Officials of the Indian space agency, ISRO,[13] provided further details for the media: a price tag of about $1.3 to $2 billion,[14] the use of India's existing medium-lift rocket (GSLV[15] Mark III), and the details of successful preparatory studies, prototyping and testing already carried out. The head of ISRO said on television that "The majority of the technologies required for this project have already been developed by my engineers." Also relevant (but not mentioned) was the agreement signed five months earlier with President Macron of France to collaborate on human spaceflight in a few specific technology areas, which may reduce some of the risks in the program [6–8].

Surprisingly for a populist politician, Modi failed to play the popular culture card that China used when it gave its Moon probes names from Chinese myths. So far India has used boring functional names for its deep space probes – its robotic Moon orbiter was called "Moon vehicle" (Chandrayaan in Sanskrit) and its probe currently orbiting Mars is called "Mars craft" (Mangalyaan in Sanskrit). The capsule to be used by India's first astronauts will be called "Gaganyaan" ("space vessel" in Sanskrit) continuing this unimaginative naming convention.

India has had its own space rocket capability since 1980, even before its development of long-range military missiles. The country has tended to avoid prestige space programs and focus on those that help provide education, monitoring of resources and weather forecasting. Its interest in human spaceflight began when China placed humans in orbit in 2003. Shortly after that India announced that it,

---

[13] ISRO: Indian Space Research Organization.

[14] Estimates being offered by Indian space agency officials six months earlier were about double these figures.

[15] GSLV: Geostationary Satellite Launch Vehicle.

too, would send humans into space, apparently determined to show that it was as technically advanced as its Asian rival. In 2009 there was even an announcement that India would send humans to the Moon by 2020. However, the funding for these plans was only sufficient for preliminary work to be undertaken, and there was no real commitment to send Indians into orbit – until now.

Despite the ISRO chief's claim that most of the technology was to-hand, India lacks a capsule capable of supporting humans in space and of surviving the high temperatures of the return to Earth. It has successfully tested a small prototype but now needs to scale that up to full size. It also lacks a rocket certified for human flights. It has successfully tested an escape system for astronauts should the rocket blow up on the pad or in flight, but that's only one element of a human-rated rocket.

Everyone agrees that the four year deadline set by Prime Minister Modi is a big ask. One solution would be to buy in technology from elsewhere, but from where? Already India has agreed to use French knowhow to help its human spaceflight ambitions in areas such as simulators and flight training. France, though, hasn't got the full portfolio of human spaceflight technology that India needs.

One definite possibility would be for India to follow China's example and buy a manned capsule from abroad to help shortcut the design process for their own craft. China purchased a Soyouz capsule and various pieces of space technology from Russia in 1994-95 which enabled them to launch the first test version of their Shenzhou spacecraft in 1999, leading to the first manned flight in 2003. That's an eight-year gap while India has given itself only four years.

Could India go the whole hog and buy or lease from Russia or China a capsule to use on this first human space mission, or get one from SpaceX or Boeing, who are developing crew transport spacecraft for NASA? This seems an unlikely prospect given the nationalistic and "go it alone" tone of Prime Minister Modi's Independence Day announcement.

India's foreign policy under Prime Minister Modi can be seen as trying to win for India a seat at the top international table. Among its Asian neighbors China alone is a permanent member of the U. N. Security Council putting *it* firmly at the table. Modi has strengthened India's military forces[16] and sought good relations with other Asian powers, including Japan, Russia and China. In terms of concrete foreign policy, one of India's top priorities is to have control of the Indian Ocean, which risks being compromised by China's actions, including the opening of China's first overseas military base in Djibouti on Africa's Indian Ocean coast and its investments in India's close neighbors such as Bangladesh and Pakistan.

Collaboration with China (short of buying the whole capsule from them) would be an interesting political move (see Fig. 12.5). It would certainly be consistent with China's invitation (open to all countries) to join in its future space station program [9]. And China certainly has technology and test facilities that would help India.

---

[16] In 2016 India became the world's fifth-largest military spender, ahead of France and the United Kingdom.

Fig. 12.5. India's Prime Minister Narendra Modi (right) meets with China's Defense Minister General Wei Fenghe August 21, 2018. The previous week, Mr. Modi announced plans for India to launch an astronaut by 2022, aiming to repeat China's 2003 feat. Modi has maintained generally friendly relations with China in spite of occasional military clashes on their disputed mutual border in the Himalayas. (Illustration from www.pmindia.gov.in.)

The economic and political rivalry between India and China is both a stumbling block and an opportunity. Collaboration in space between the two Asian powers would send a message to the traditional space powers in the West and Russia that the game has changed. For example, China plus India could demand equal partner status with the United States in the Lunar Gateway program.

However, part of the rationale for an Indian human spaceflight program is to catch up with China. Being seen to rely on China for key technologies or facilities would confirm that India lags behind in space. It would take a brave Indian politician to admit that India needed the help of China to undertake such a high profile space mission. But if India wants to be at the space "top table" a liaison with China may be a sure and fast way to get there.

India has long-standing ties with Russia in defense dating from when the West embargoed military sales to India because of its development of nuclear weapons. And there have been close ties in space technology with Russia, including the transfer of crucial rocket technology to India. Things turned sour in 2013 when Russia pulled out of the *Chandrayaan -2* joint robotic mission to the Moon. Russia was to have provided the craft that lands on the Moon while India would provide a rover carried in the lander and a spacecraft that orbits the Moon. The agreement had been signed in 2007, with the launch planned for 2015, so this late change of heart by Russia had a major impact, caused by a failure of the technology to be supplied by Russia in another of its space missions.[17] Eventually India decided to go it alone but has had to delay the launch until 2019.

Russia certainly has technology, facilities and knowhow that could help India in its first human spaceflight project. Given what happened with *Chandrayaan-2*, India will presumably be nervous about relying on Russia for any crucial elements of the 2022 mission, but some assistance from Russia might well be possible – technically and politically.

The final major potential collaborator is of course the United States. U. S. collaboration with India on exploration of the Moon was highlighted by NASA administrator Jim Bridenstine in August 2018 when he cited the discovery of water on the Moon's surface by a U. S. instrument on India's *Chandrayaan-1* probe orbiting above. Bridenstine spoke of "hundreds of billions of tons of water-ice on the surface of the Moon" as one of the keys to a long-term sustainable exploration program. Mind you, we know (Chapter 10) that NASA's ability to create a *sustainable* Moon landing program is in doubt, but NASA's chief clearly values the outcome of the U. S.-India cooperation [10].

Despite India continuing to import military equipment from Russia, Modi has however also significantly increased his arms imports from the United States,[18] thereby avoiding dependence on a single source of weaponry. An added feature of this relationship is the close personal ties between India and the United States in high tech, especially in computing and software[19] and also in space. The most famous space example is probably NASA astronaut Sunita Williams (see Fig. 12.6), whose father was from India and who was the commander of the ISS in 2012.

Incidentally, India's space agency, ISRO, has appointed a woman to manage the human spaceflight project, V R Lalithambika, who has a background in the development of India's space rockets, most recently as deputy director of ISRO's VSSC[20] launch center on the coast at the southern tip of India. Could the presence of such two high-powered women allow a sensible collaboration to be worked out? A pragmatic arrangement between the United States and India could well be

---

[17] The Fobos-Grunt mission to Mars.

[18] India imports from U.S. companies have gone from close to zero to more than $15 billion worth since 2007.

[19] For example Satya Nadella (Microsoft CEO), Padmasree Warrior (Cisco CTO) and Shantanu Narayen (Adobe CEO).

[20] VSSC = Vikram Sarabhai Space Centre. Indian scientist Dr. Vikram Sarabhai (1919-1971) is widely regarded as the father of the Indian space program.

Fig. 12.6. ISS Commander Sunita Williams briefing the media from space in October 2012. As a highly successful NASA astronaut with Indian heritage she has done much to enhance the image of human spaceflight in India and to strengthen U. S.-India ties. (Illustration courtesy of NASA TV.)

the route to ensuring that India meets its 2022 deadline for launching its first astronaut.

If India succeeds in launching astronauts by 2022, that is still a long way from being able to place them on the Moon. It might however give India something with which to negotiate a role in the Moon-landing plans of a more advanced space country such as the United States, Russia or China. The choice of partner would probably be dictated by the political scene at the time. For example, will the United States be at loggerheads with the rest of the world because of tariffs introduced by President Trump? In this case India might team up with China to make the point that Asia is the most important economic dynamo of the world in the 21st century. Or for example if American private sector Moon-landing plans end up being in competition with those of NASA, India might help provide international legitimacy to the U. S. company.

With so many personal links between India and the United States in high tech sectors, it seems eminently plausible for the two countries to work together on future human spaceflight missions.

References

1. "Russia's Pivot to Asia", *The Economist* Banyan Column, Nov. 26 2016 online at https://www.economist.com/asia/2016/11/26/russias-pivot-to-asia (Accessed Oct. 8, 2018).
2. Norris, P., *Watching Earth from Space*, Springer Praxis (Chichester, UK), 2010, pp252-254.
3. Hendrickx B, *Russia's evolving rocket plans*, Space Review, Sept. 5, 2017 see http://www.thespacereview.com/article/3321/1 (Accessed Oct. 8, 2018).
4. Zak, A., "Russia Is Now Working on a Super Heavy Rocket of Its Own", *Popular Mechanics*, Feb. 8, 2018, online at https://www.popularmechanics.com/space/rockets/a16761777/russia-super-heavy-rocket/ (Accessed Oct. 8, 2018).
5. http://planet4589.org/space/papers/Edge.pdf (Accessed Oct. 8, 2018).
6. Kumar, C., *India to launch first manned space mission by 2022: PM Modi*, The Times of India, online at https://timesofindia.indiatimes.com/india/india-to-launch-first-manned-space-mission-by-2022-pm-modi/articleshow/65410373.cms (Accessed Oct. 8, 2018).
7. Bagla, P., *Modi-Macron Space Plan: India Partners France In Bold Vision For Space*, NDTV, online at https://www.ndtv.com/india-news/modi-macron-space-plan-india-partners-france-in-bold-vision-for-space-1822293 (Accessed Oct. 8, 2018).
8. Mouriaux, P. F., "L'Inde se (re)positionne pour 2022," *Air & Cosmos,* August 31, 2018, p39.
9. China National Space Administration press release, *China welcomes all UN member states to jointly utilize its space station*, May 30. 2018, see http://www.cnsa.gov.cn/n6443408/n6465652/n6465653/c6801729/content.html (Accessed Oct. 8, 2018).
10. Gorman, S., Rabiee M, *NASA chief excited about prospects for exploiting water on the moon*, Reuters, Aug. 22, 2018, online at https://www.reuters.com/article/us-nasa-bridenstine/nasa-chief-excited-about-prospects-for-exploiting-water-on-the-moon-idUSKCN1L7062 (Accessed Oct. 8, 2018).
11. Day, D, "There are no Russians there", *The Space Review*, Nov. 26, 2018 online at http://www.thespacereview.com/article/3611/1 accessed Nov. 30, 2018.

13

Conclusions

For the first time since the 1960s there are realistic schedules for people to once again walk on the Moon. Possibly by 2024 and almost certainly by the end of the 2020s a U. S. astronaut will follow in the footsteps of Gene Cernan and relieve him of the "last man on the Moon" title. SpaceX is the bookmaker's favorite to provide the technology to reach the Moon first, although there is a good chance that it will be jointly with NASA.

If SpaceX slips up (and with an owner who admits to working 120 hours a week and to taking Ambien to help sleep [1], there must be a risk) there are other serious candidates warming up. It might be the early to mid-2030s, but in the United States, Blue Origin might have deployed a super-big rocket, or NASA might get the funding it needs to complete its own super-rocket, and outside the United States, China seems committed to following its robotic lunar missions with a manned one. Let's look at this in a bit more detail.

As described in earlier chapters, Apollo happened because its success was made a national strategic priority with a budget to match for several years. From a technical point of view the key development was the Saturn V heavy-lift rocket. But in addition, NASA's effective management of the program was crucial in sticking to President Kennedy's 1961 commitment: *Before this decade is out, of landing a man on the Moon and returning him safely to Earth*. It was an amazing achievement!

The difficulty of landing men on the Moon was highlighted by the fact that the Soviets tried to beat the Americans but in fact failed to get there at all. Soviet rocket technology was first class, as witnessed by the fact that the rockets it developed in the 1960s are still in widespread use today. The Soyuz and Proton rockets are the prime examples. But they failed to develop a heavy-lift rocket on the scale of the Saturn V – though they certainly tried (see, for example, Figs. 8.14 and 8.15).

The bottom line was that the Soviet way of managing space programs was simply not up to the task of a Moon program. They never provided the funding to go with various commitments to a manned Moon landing, they kept interrupting the program to focus on short-term priorities and they frequently changed designs and suppliers. But these management follies were an inevitable feature of the communist way of running the country. Russian engineers and managers could work around the system sufficiently to launch the first robotic satellite in 1957, the first human in 1961 and many other space "firsts," but sending men to the Moon was just too complicated for a work around.

Still, in the 50 years since *Apollo 11,* surely some nation should have placed humans on the Moon? We have seen in earlier chapters (especially Chapter 9) that getting people to the Moon's surface and back requires a rocket that is about five times bigger than needed for any other purpose. Of the hundred or so space launches each year (satellites for TV broadcasting, for satnav, for weather forecasting, for science, etc.) none requires a Saturn V-type rocket. So you have to justify the cost of developing a super-rocket solely on the basis of a Moon trip, making that trip extremely expensive.

Instead of using a single very large rocket, there have been suggestions to launch smaller rockets several times and assemble the Moon ship in space. We have learned the hard way with the ISS that assembling a complex structure in space is difficult, time-consuming and very expensive. The ISS has cost about as much as the Apollo program (both about $100 billion in today's money). So this approach is also very costly and therefore difficult to justify.

Going to the Moon and back without landing on the surface can be done – just barely – using the largest of rockets used for other purposes. It's the extra power needed to land without parachutes (no atmosphere on the Moon) and then take off again and get up to orbit speed that makes a super-rocket essential.

A U. S. company, SpaceX has now produced and launched in February 2018 the Falcon Heavy rocket that is about half the power of Apollo's Saturn V and is developing an even more powerful rocket and spaceship called Super Heavy and Starship (Fig. 13.1) that will exceed the Saturn V's performance. The owner (Elon Musk)'s vision of traveling to Mars is his justification for spending the perhaps $5 billion needed for the development of the super-rocket.

His skill at developing rockets for a fraction of the cost of NASA has already been demonstrated in the form of his mainstream and best-selling Falcon 9 rocket, whose low price has forced all rocket suppliers to reduce their prices significantly, and by the Falcon Heavy, which is the world's most powerful rocket at the moment and was developed without government funding. This gives him a revenue stream that underpins the next round of development and makes his plans credible. SpaceX will pass an important milestone in 2019 with its first launch of a human crew on its Falcon 9 rocket. The crew will be transported to and from the ISS in orbit about 250 miles (400 km) above Earth.

212   **Conclusions**

Fig. 13.1. Artist's rendering of SpaceX's fully reusable Super Heavy rocket and Starship spacecraft as the two separate. The large number of portholes indicates the intent to carry a crew of about 40. The fins provide landing legs as well as in-atmosphere control. (Illustration courtesy of SpaceX. Used with permission.)

Super Heavy / Starship is being designed especially for a Mars mission, but SpaceX reckons it can also be used for human missions to the surface of the Moon. They have sold several seats on a Starship flight that will go around the Moon and straight back to Earth (not landing on the Moon) by about 2023. (Musk admits that this date is an optimistic one, but claims it is doable.)

However, the sobering fact is that SpaceX has a key-person dependency risk in the form of its owner, Elon Musk, whose death or incapacity would jeopardize its Moon and Mars plans.

Another U. S. company owned by a billionaire, Blue Origin, is aiming to establish human outposts on the Moon and is developing rockets that may eventually be powerful enough for that job. Blue Origin's owner, Jeff Bezos, is the world's richest man and thus has enough money to be able to fund the rocket development at his own discretion. His suborbital rocket has been demonstrated, and his first rocket capable of achieving orbit, called New Glenn, is far enough along in development to be credible, with a first orbital mission scheduled for 2020. New Glenn will be almost as powerful as the Falcon Heavy and about 40 percent of the power you would need for a manned lunar landing.

Bezos has stated that his objective is to establish industrial facilities on the Moon, and he has hinted that a rocket big enough for the task is in the pipeline. Its name will be New Armstrong. New Glenn is being named after the first American

Conclusions 213

to orbit the Earth, New Armstrong presumably will be named after the first American to step onto the Moon, which indicates what the New Armstrong rocket will be capable of.

Like SpaceX, Blue Origin has a key-person dependency risk in the form of Bezos himself. Would a successor have the vision of industrializing the Moon and be ready to invest a billion dollars a year to achieve it?

NASA did it in the 1960s, so it would be foolish to ignore its plans, especially as it has spent $23 billion in the decade up to 2018 on developing a super-rocket, the spacecraft it would carry to the Moon and the ground facilities to support them. The Space Launch System super-rocket continues to be funded at around $2 billion a year, and is due for its first launch in 2020 (see Fig. 13.2). There are to be three versions of the Space Launch System, the 2020 version being roughly as powerful as SpaceX's existing Falcon Heavy. The second (intermediate) version is due to be launched in about 2024, with the final version (the version you would need to land humans on the Moon with a single launch) available five to ten years after that.

NASA's costs are sky high, and it will probably need $15 billion or more to meet the 2024 deadline it has been given by the Trump Administration. There must be some doubt that Congress will approve that sort of funding – especially as there are private sector alternatives becoming available at a fraction of the cost.

A program in which NASA's Space Launch System is sidelined and replaced by SpaceX and/or Blue Origin rockets in the early 2020s looks like a good bet.

Fig. 13.2. Artist's rendering of the initial crew vehicle version of NASA's Space Launch System taking off from Cape Kennedy. This version of the Space Launch System will be only slightly more powerful than the existing SpaceX Falcon Heavy. More powerful versions of SLS are planned. (Illustration courtesy of NASA.)

214 Conclusions

In summary, by the mid-2020s at least one U. S. rocket in the Saturn V class should be available, and by 2030 perhaps three different ones. This would open the door for American astronauts to return to the surface of the Moon.

We saw in Chapter 12 that India, Russia and other countries have no serious plans for sending humans to the Moon. India is planning to launch an astronaut by 2022, but has no credible plans for a human mission to land on the Moon.

Russia has been trying without success to develop new rockets for 15 years, so statements about them building a super-rocket lack credibility for the foreseeable future. In fact, there is little sign that Russia has learned the lesson of the 1960s that efficient management from top to bottom is key to success when considering such a complex mission as landing humans on the Moon.

Europe, Canada and Japan have no plans to develop rockets in the Saturn V class, and will probably collaborate with the United States in its Moon program.

China is a different matter, as it has indicated that it is considering sending humans to the Moon. It has begun by targeting the Moon with robotic probes in innovative ways, such as the 2018/2019 *Chang'e 4* rover on the hidden far side of the Moon and the 2019/2020 *Chang'e 5* probe to return 5 pounds l (2 kg) of Moon rock to Earth for analysis.

In parallel with that, China has revealed long term plans [2] that include a heavy-lift rocket for which the prototype of the engine is nearly ready and a new spacecraft large enough to support a mission to the Moon's surface. The year 2030 is the likely date by which the heavy-lift rocket will be ready for use. India's recently announced plans to launch an astronaut by 2022 may encourage China to continue with its 2030 plan in order to convincingly maintain its space leadership in Asia.

Thus China looks like it will be sending men to the Moon a few years after the United States. But if there are any delays in the U. S. plans, then China will be positioned for a major publicity coup, with Gene Cernan's "last man on the Moon" title passing to a Chinese astronaut.

We saw in the 1960s there's nothing like a good competition to get things moving. In the 2020s the competitive environment will be weaker than sixty years earlier, but there is nevertheless sufficient motivation to demonstrate their national capability in the United States and China that a manned Moon mission looks likelier now that at any time since *Apollo 17,* perhaps by 2024 and very probably by 2030.

References

1. Gelles, D., et al, "Elon Musk Details 'Excruciating' Personal Toll of Tesla Turmoil," *New York Times,* Aug. 16, 2018, online at https://www.nytimes.com/2018/08/16/business/elon-musk-interview-tesla.html (Accessed Oct. 12, 2018).
2. Jones, A., "Chinese space program insights emerge from National People's Congress," *Space News,* April 2, 2018 online at: http://spacenews.com/chinese-space-program-insights-emerge-from-national-peoples-congress/.

Glossary

Airlock   Soviet spacecraft required an airlock between the living area and a chamber from which a cosmonaut could exit to perform a space walk. The U. S. Apollo and Gemini spacecraft had no airlock or exit chamber, depressurizing the whole cabin and exiting direct from there. Equipment inside the U. S. craft was designed to work in vacuum, while that in the Soviet craft could not survive in a vacuum. The U. S. craft would of course be re-pressurized once the astronaut returned from the space walk and the hatch was closed.

Ascent Stage   The Apollo Lunar Module was comprised of a Descent Stage and an Ascent Stage. The Ascent Stage comprised the living cabin, the ascent engine plus equipment and resources necessary for the journey from the Moon's surface to the Command and Service Module in orbit above. See Figs. 4.6 and 4.12.

Atlas rocket   The Atlas rocket family started out as an Intercontinental Ballistic Missile (ICBM). Atlas rockets launched the Mercury capsule into orbit, including the first American to orbit Earth, John Glenn, and three other astronauts, Scott Carpenter, Wally Schirra and Gordon Cooper. More powerful versions of the Atlas also launched the Ranger (hard-landing) and Surveyor (soft landing) Moon probes. Surveyor used the version with the Centaur upper stage fueled by liquid hydrogen and liquid oxygen. The success of the Centaur upper stage led to the same fuel combination being used for the second and third stages of the Saturn V. The Atlas continues to be a workhorse launch vehicle of the U. S. Department of Defense and (to a lesser extent) of NASA.

Ballistic   A rocket or missile is said to be in ballistic flight when its rocket engines are not in use. Its trajectory is then determined only by forces such as gravity and air drag.

BE-4 rocket engine   The BE-4 engine is under development by Blue Origin, designed to be used in its New Glenn rocket. Seven BE-4 engines will power the first stage. The BE-4 has also been sold to the United Launch Alliance to be used in its Vulcan rocket. The BE-4 uses liquid methane and liquid oxygen as fuel, developing a thrust of about 2,400 kN at sea level, which is about five times that of engine used in the New Shepard rocket.

Blue Origin Amazon founder and Chief Executive Officer Jeff Bezos has been funding his Blue Origin rocket company to the tune of $1 billion a year. The rockets are reusable and intended to fly humans to space. See BE-4, New Shepard and New Glenn elsewhere in this glossary, and Chapter 10.

Canadarm The robotic arm attached to the exterior of the ISS is used to move heavy or bulky objects around the outside of the station, controlled by an astronaut inside it. It was provided by Canada, hence its name.

Centaur rocket stage The Centaur was the first operational rocket stage fueled by liquid hydrogen and liquid oxygen and built by General Dynamics (now United Launch Alliance). It is comprised of one or two RL-10 engines built by Pratt & Whitney (now Aerojet Rocketdyne). Its first successful flight was in 1962, and variants of it are still in use today.

Chandrayaan India's series of robotic Moon probes are called Chandrayaan ("Moon vehicle" in Sanskrit). *Chandrayaan-1* was launched into orbit around the Moon in 2008 at an altitude varying between 90 and 170 miles (150 and 270 km). It stopped sending back information after ten months. *Chandrayaan-2* is scheduled for launch in 2019 and will include a lander with a rover, as well as a probe in orbit around the Moon – see Chapter 12.

Chang'e China's series of robotic Moon probes are named after the Moon goddess. *Chang'e-1* was launched in 2007 and orbited the Moon for 1 year 4 months. The images it sent back were used to make a detailed map of the Moon. *Chang'e-2* was an identical probe launched in 2010. Having spent 8 months orbiting the Moon, and 8 months at the L2 Lagrange point 1 million miles from Earth, where Earth and Sun gravity are equal, *Chang'e-2* headed out into the Solar System to intercept the asteroid Toutatis. It passed within 2 miles (3.2 km) of Toutatis in December 2012, whizzing by at a speed of 6.7 miles/second (10.5 km/sec, 24,000 mph) and capturing detailed images of the asteroid's surface. *Chang'e-3* was launched in 2013, orbited the Moon for a week, then landed in Mare Imbrium on the front face of the Moon. It deployed a rover called Yutu or jade rabbit (who in legend accompanied the Goddess Chang'e to the Moon). *Chang'e-4* is similar to *Chang'e-3* and in early 2019 landed and deployed a rover on the lunar far side. The *Queqiao* (*Magpie Bridge*) satellite was launched in May 2018 and is located above and behind the Moon so that it can relay radio signals from *Chang'e-4* and its Yutu rover. *Chang'e-5* is scheduled to return samples from the Moon's surface to Earth in about 2020. See also Chapter 11.

Circumlunar A space mission that goes around the Moon and straight back to Earth (without going into orbit around the Moon) is called circumlunar. The most famous such mission was *Apollo 13* in 1970, which had to return as fast as possible to Earth after an oxygen tank exploded in the Service Module.

Clementine NASA's first mission to the Moon after *Apollo 17* was the 1994 Clementine robotic probe developed jointly with the Department of Defense. Clementine spent three months orbiting the Moon, taking detailed images and scientific measurements, then was due to head for the asteroid 1620 Geographos when a failed thruster left it in Earth orbit instead, ending its mission a month later.

Command Module The Apollo Command Module was the living quarters for the three astronauts during the mission to the Moon that could last up to 12 days – see Chapter 3.

Command Module Pilot The Command Module Pilot remained in orbit around the Moon while his two companions landed on the surface in the Lunar Module.

Communications satellite Satellite that links two Earth-bound terminals or that broadcasts TV to many terminals. They are particularly useful for linking or broadcasting to parts of Earth that are otherwise difficult to reach such as islands, remote regions, ships at sea and aircraft in flight. Many are in geostationary orbit, about 22,000 miles (36,000 km) high, where they appear stationary from Earth's surface.

CORONA spy satellite The Central intelligence Agency with technical support from the Department of Defense launched nearly 150 CORONA satellites in the 1960s to take images of the Soviet Union and other countries of interest. The satellites returned the exposed film to Earth in one or more re-entry capsules. About 70 percent of the satellites returned usable imagery. CORONA was designed to take wide-area pictures rather than highly detailed ones – the Air Force Gambit satellites performed that role. The Soviet Zenit satellites worked on a similar principle to CORONA.

Cosmodrome Soviet and Russian space launch sites are called cosmodromes – see Fig. 12.1.

Cosmonaut Russian astronauts are called cosmonauts.

Descent Stage The Apollo Lunar Module was comprised of a Descent Stage and an Ascent Stage. The Descent Stage was comprised of the descent engine with its fuel tanks, the landing structure (with spider's legs), scientific equipment to be left on the Moon's surface and (*Apollo 14, 15, 16* and *17*) the lunar rover. See Fig. 4.11.

Docking Spacecraft that join up in space so that they become a single vehicle are said to "dock." Examples include the Command and Service Module docking with the Lunar Module and extracting it from the Saturn V third stage on the way to the Moon, and the Command and Service Module docking with the Ascent Stage of the Lunar Module when it returns from the Moon's surface.

Earth orbit An object in space regularly circling another object is said to be in orbit around it. For a manmade object to orbit Earth it needs to get above the bulk of the atmosphere, i.e., about 100 miles up, and to travel at 17,500 mph (28,000 kmph) or more.

European Space Agency (ESA) ESA undertakes space missions on behalf of its 22 member countries. Increasingly ESA also undertakes space missions for the European Union. (About a quarter of ESA's 2018 budget of $6.4 billion – 5.6 billion Euros – was provided by the European Union.) A recent highlight was the two-year-long rendezvous of ESA's Rosetta probe with a comet. ESA is a partner in the ISS, as well as the Hubble Space Telescope and its planned successor, the James Webb Telescope, and is supplying the Service Module for NASA's Orion capsule (see Chapter 12).

Explorer satellites Early NASA scientific satellites were given the general purpose title "Explorer" followed by a series number. The large number of these satellites made it impossible to tell their function from the name, so most were also given informal mission-related names, for example Explorer-51, launched in 1973 was called Atmosphere Explorer-C.

218   **Glossary**

Extra-Vehicular Activity (EVA)   Exiting from a spacecraft in space is officially called an extra-vehicular activity (EVA) by NASA. When the spacecraft is in orbit, the EVA is popularly known as a "space walk". On the surface of the Moon the EVA is sometimes referred to as a "Moon walk."

F-1 engine   Developed by Rocketdyne in close collaboration with NASA's Marshall Space Flight Center, five of the F-1 engines were used in the first stage of the Saturn V rocket. The propellant used was a refined form of kerosene and liquid oxygen. See Chapter 2.

Falcon 9   Falcon 9 is SpaceX's current workhorse rocket. The first stage is comprised of 9 Merlin engines burning rocket-grade kerosene and liquid oxygen. The second stage contains a single modified Merlin engine. Since 2015, having separated from the second stage, the first stage can make a controlled return to the ground, either to the launch site or to a floating platform, so that it can be reused. Falcon 9 can lift about 23 tons to low Earth orbit, although that has to be reduced (by as much as a third) if the first stage is to be recovered. There have been nearly 20 launches per year in 2017 and 2018, with about three-quarters involving recovery of the first stage. The first stage of the Block 5 version introduced in 2018 can be re-flown up to ten times (the previous version could only re-fly once).

Falcon Heavy   Strapping three Falcon 9 rockets together gives the Falcon Heavy, which can lift 64 tons to low Earth orbit. Development took longer than predicted by SpaceX, leading to the first launch in 2018. SpaceX had planned to upgrade the Falcon Heavy to carry humans but now says that no significant further upgrades to Falcon Heavy will take place. Instead, SpaceX is developing the Super Heavy and Starship rocket and spaceship. See Chapter 10.

Gemini spacecraft   The two-person Gemini capsule was an intermediate development between the single-person Mercury and the three-person Apollo capsules. Ten Gemini spacecraft were launched in 1965-66, testing techniques such as docking, rendezvous in orbit, space walks (EVA) and two-week long stays in space, all essential for the Apollo Moon missions. Gemini also gave many Apollo astronauts their first space experience. The prime contractor was McDonnell Aircraft Corporation, which had also been the prime for Mercury.

Gravity field   The force of gravity exerted by Earth would be identical wherever you are on the surface if Earth were a perfect and uniform sphere. Earth, however, has bulges and regions of heavier material that change the force of gravity. The sum total of all these variations is called the gravity field. The Moon's gravity field is even more uneven than Earth's.

Heavy-lift launcher   Heavy-lift is a loose way to describe the most powerful rockets. As more powerful rockets become available, what was a "heavy-lift" rocket becomes a "medium lift" one.

Intercontinental Ballistic Missile (ICBM)   Missiles that can hit a target beyond about 3,500 miles (5,500 km) are said to be intercontinental in their range. A missile is "ballistic" if it coasts for most of the journey, having been boosted up to high speed initially by its rocket motor.

International Space Station (ISS) Astronauts undertake scientific and technical tasks in the ISS orbiting Earth about 250 miles (400 km) out in space. It took 13 years to assemble all the parts, which in total weigh more than 400 tons. It has been continuously occupied since November 2, 2000, and is expected to continue operation until at least 2025. Continued funding of the station makes it difficult for NASA and the other partners to fund a human Moon landing mission – see Chapters 10 and 12.

Indian Space Research Organization (ISRO) ISRO is the civil space agency for the government of India, and reports via the Department of Space to the Prime Minister. Its annual budget is about $1.5 billion and it has about 15,000 staff.

J-2 engine Five of Rocketdyne's J-2 engines powered the second stage of the Saturn V rocket, and one powered the third stage. Like the RL-10 engine that powered the Centaur rocket stage, the J-2 used liquid hydrogen and liquid oxygen as propellants, but was ten times more powerful – see Chapter 2.

L1 spacecraft The Soviet L1 spacecraft was intended to carry two cosmonauts around the Moon and then straight back to Earth – a circumlunar mission. Derived from the Soyuz spacecraft and launched on the Proton (UR-500) launcher, two test versions of the L1 completed the circumlunar trip in October and November 1968, some carrying small animals and plants, but no human passengers ever made it. After the success of NASA's *Apollo 8* in December 1968 carrying three astronauts into orbit around the Moon, the L1 program was quietly ended without its existence ever being made public – see Chapter 8.

L3 Spacecraft The L3 was the manned payload to be launched towards the Moon by the Soviet N1 rocket. It consisted of two main parts: the LOK mother ship that carried the cosmonauts to and from Earth and the LK spacecraft that would land on the Moon's surface.

Launch Escape System Above the Apollo spacecraft on top of the Saturn V rocket was a small rocket assembly called the Launch Escape System. If the Apollo spacecraft was in danger during the first few minutes of flight this system would be fired and would lift the spacecraft clear of the Saturn V. See Fig. 3.3.

LK spacecraft The LK spacecraft was the Soviet equivalent of the Apollo Lunar Module. At about 5½ tons it weighted about a third of the Lunar Module (Fig. 8.13) and was designed to carry a single cosmonaut from the Soyuz-type LOK capsule in orbit around the Moon down to the Moon's surface and back. Prototypes, called T2K, were successfully tested in Earth orbit three times in 1970-71. A number of engineering models of the LK remain (Fig. 8.12). See Chapter 8.

LOK Spacecraft The LOK was a variant of the Soyuz capsule and was designed to carry two cosmonauts to the Moon and back. It was intended to go into orbit around the Moon using its built-in rocket engines with the LK lunar lander attached, then to dock with the LK when it returned from the Moon's surface, and then leave lunar orbit and return to Earth.

Long March rockets China's main space launch vehicles are usually given the name Long March (LM) in the West (Changzheng in Chinese pinyin). Various versions of the LM-2, LM-3 and LM-4 rockets have been in use since the 1980s using mainly storable propellants. (The main exception is that the upper stage of some versions of the LM-3 uses liquid hydrogen and liquid oxygen.) The new LM-5, LM-6 and LM-7 rockets are discussed in Chapter 11.

Longjiang Two Chinese robotic microsatellites were launched to the Moon in May 2018 as secondary passengers on the launcher that carried the Queqiao relay satellite to the Moon. *Longjiang-1* suffered a failure, but *Longjiang-2* successfully entered lunar orbit and sent back images taken by a camera supplied by Saudi Arabia, including at least one Earthrise picture. The main mission of the Longjiang satellites is radio astronomy, to take advantage of being shielded from Earth's radio signals when behind the Moon.

Lunar Gateway NASA plans to create an inhabited space laboratory in orbit around the Moon, frequently referred to as a Lunar Gateway. See Chapter 10.

Lunar Module The Apollo Lunar Module was the spacecraft that carried two astronauts to the surface of the Moon and then back to the Command and Service Module that waited in orbit around the Moon. See Chapter 4 and Fig. 3.10.

Lunar Module Pilot The two-man crew of a Lunar Module was comprised of the Commander and the Lunar Module Pilot.

Lunar Orbiter Five robotic Lunar Orbiter spacecraft launched in 1966 and 1967 took detailed images of the Moon in preparation for the Apollo missions. They also enabled analysts back on Earth to examine details of the Moon's gravity field by observing how the trajectories of the Lunar Orbiters varied as they circled the Moon.

Lunar Reconnaissance Orbiter This was a NASA robotic probe launched in 2009 to provide detailed imagery of the Moon and to analyze the chemistry of the lunar surface. Its eccentric orbit typically brings it to within 12 miles (20 km) of the surface, with a high point of about 100 miles (160 km). See Fig. 7.2.

Lunokhod The Soviet Union had two successful robotic Moon rover probes, Lunokhod-1 in 1970-71, carried to the Moon on *Luna-17,* and Lunokhod-2 in 1973 *(Luna 21).* An earlier Lunokhod was destroyed during launch in February 1969. See Chapter 8.

Mascons Areas on the Moon where the pull of gravity is greater than average, typically associated with the circular seas or mares – see Chapter 4.

Mercury spacecraft The one-person Mercury spacecraft, built by McDonnell Aircraft, carried America's first astronauts into space in the early 1960s. There were six manned Mercury flights starting with the suborbital flight of Alan Shepard in May 1961 (see Chapter 1). The first American to orbit the Earth, John Glenn, was in *Mercury 3* in February 1962. Two 1961 test flights carried a chimpanzee – one suborbital in January the other into orbit in November.

Merlin engine The SpaceX Falcon 9 rocket is powered by nine SpaceX Merlin engines. The propellant is rocket grade kerosene and liquid oxygen. See Chapter 10.

Mission Control Once in space, the Apollo missions were managed from Apollo Mission Control in the Manned Spacecraft Center in Houston, Texas. Working a shift system, teams of experts provided support to the crew in space, aided by further teams in industry and other NASA centers around the country.

Moore's law The seemingly inexorable improvement in electronics since the 1960s follows a pattern first articulated by Gordon Moore of Intel – hence the name Moore's law. See Chapter 9 and Fig. 9.2.

N1 rocket The nearest Soviet equivalent of the Saturn V rocket was the N1. Although heavier than the Saturn V, the N1 could only carry a payload of just under 100 tons into orbit compared to Saturn V's 130 tons. There were four failed launch attempts: on February 21, 1969, the engines shut down 70 seconds into the flight.; on July 3, 1969, the rocket rose 700 feet (200 m) high then fell back onto the launch pad and exploded; on June 27, 1971, the engines shut down after 51 seconds; and on November 23, 1972, the rocket exploded after 104 secs. See Chapter 8 and Figs. 8.14 and 8.15.

NACA The National Advisory Committee for Aeronautics (NACA) was founded in 1915 to undertake aviation research in the United States and was merged into the newly created NASA in 1958.

National Aeronautics and Space Administration (NASA) NASA is responsible for America's civilian space program plus research in aviation and aerospace. It was created by President Eisenhower in 1958 to provide a federal agency to deal with non-military space matters.

New Shepard The New Shepard rocket developed by Blue Origin is able to carry up to six passengers to an altitude of about 62 miles (100 km) and then return to Earth to be reused. The passenger capsule descends on a parachute with small rockets ensuring a smooth touchdown. The rocket lands upright, using its engines to control its descent. Test flights have been undertaken since 2015 and first operational flights are planned for 2019. The rocket is named after the first American to make a suborbital flight beyond 100 kilometer altitude, Alan Shepard in the Mercury capsule.

Orbit rendezvous For two objects to remain close together for an extended period of time (i.e., to rendezvous) in space their orbits have to be identical. This means that when they are together their speed in every direction (up/down, forward/back, sideways) has to be identical. To avoid using unnecessary fuel, two objects in space need to gradually align their orbits to steadily reduce their separation.

Orbital bombardment system The Soviet Union investigated the possibility of placing a nuclear-armed missile in orbit able to be commanded to hit a target on Earth if and when required. The concept was called the GR-1 orbital bombardment system. It was part of the justification for developing the N1 rocket, but was terminated in 1965 due to lack of military interest.

Orion The crewed spacecraft to be carried by NASA's new Space Launch System is called Orion – being developed by Lockheed Martin for NASA and incorporating a Service Module built by Airbus for the European Space Agency. See Chapter 10.

Payload The cargo and/or passengers carried by a rocket is called the payload. The name derives from the fee paid to carry the cargo or passengers to space.

Proton (UR-500) rocket The Proton is the most powerful Russian rocket currently in use (see Table 9.1). Originally called the UR-500, more than 400 Protons have been launched since the first one in 1965.

R-7 rocket Developed by Sergei Korolev, the R-7 was the world's first ICBM and was the basis for the Vostok rocket that launched the first artificial satellite, *Sputnik-1,* in 1957 and the Soyuz rockets still in use today – see Fig. 8.5. Because of its cryogenic propellant (liquid oxygen), the R-7 was poorly suited to the role of military missile, and only a handful were ever deployed.

Radar A radar (radio detection and ranging) transmits a radio signal and detects the (usually much weaker) signal echoed back by a target. This allows the radar to work out the direction, distance, velocity and even the shape of the target. Civil aircraft, ships and spacecraft usually contain a device called a transponder that boosts the power of the echo so that it can be detected at a much greater distance – see Chapter 5.

Raptor engine SpaceX is developing the Raptor engine to power its Super Heavy / Starship rocket. Using liquid methane and liquid oxygen as propellants, Raptor will have more than double the thrust of SpaceX's existing Merlin engine. See Chapter 10.

Reconnaissance satellite The original motivation for an American space program was to get images of the Soviet Union. Satellites that contain a camera to collect images of Earth below are labeled as "reconnaissance" or "Earth observation" or "remote sensing" or sometimes "spy." The overhead imagery (some taken by aircraft, some by satellites) of the world on Google Earth and its imitators has familiarized us with the concept. The American CORONA and Soviet Zenit satellites of the 1960s provided accurate information about missile deployment of both the superpowers, enabling them to trust each other enough to sign up to the Strategic Arms Limitation Treaty that drastically slowed the escalation of the nuclear arms race. "Trust, but verify" as President Ronald Reagan famously said – reconnaissance satellites do most of the "verify" bit.

Re-entry When a spacecraft is returning to Earth from space, it meets strong resistance when it encounters the atmosphere. The air resistance slows the spacecraft down, and the temperature increases. The atmosphere is detectable out to about 400 miles (600 km), but it provides significant resistance to a returning spacecraft from only about 100 miles (160 km) up, that point being considered the re-entry point. It varies in altitude over the course of a day and due to changes in the intensity of radiation and charged particles from the Sun (notably over the eleven-year solar cycle).

Robotic probe An unmanned space probe is said to be "robotic." It usually contains a computer that controls its actions, and this can in turn usually be reprogrammed from Earth.

Rocket engine A rocket engine works on the principle that by pushing a gas out one end, the engine will move in the other direction – in accordance with Newton's Third Law of Motion (for every action there is an equal and opposite reaction). The gas can be simply stored and then ejected or can be the result of a chemical reaction (for example kerosene burning oxygen). Rocket engines using chemical fuels were invented more than a thousand years ago in China (steam rocket engines seem to have been known to the Greeks and Romans). Unlike internal combustion or jet engines, rocket engines work in a vacuum, hence their use in space.

Roscosmos Roscosmos is a state-owned corporation responsible for Russia's civil space programs. It was created in 1992 as a government agency and given its corporation status in 2015, when it absorbed much of the state-owned space industry.

Rover Early space probes that landed on other Solar System bodies such as the Moon and Mars stayed stationary at their landing spot. Later, a wheeled vehicle (often called a "rover") was added to the probe so that some at least of the scientific instruments could travel across the surface of the body. Typically part of the vehicle remains stationary, acting as a communication hub for the rover and performing scientific experiments that were too complex to be carried out on the rover. A rover was also carried by the later Apollo missions to the Moon (*Apollos 14* through *17*) to carry the astronauts, their equipment and the rocks and dust they collected.

RS-25 engine Built by Aerojet Rocketdyne the RS-25 is a reusable rocket motor powered by liquid hydrogen and liquid oxygen, three of which are comprised of the main engine of the space shuttle. A total of 46 reusable RS-25 engines were flown during the space shuttle program. Those that remain in the NASA inventory are now slated to be used once only in the Space Launch System currently under development – see Chapter 10.

Satellite An object circling a body through the force of gravity is said to be a satellite of the body. The Moon is the only natural satellite of Earth. Manmade objects orbiting the Earth are said to be "artificial" satellites.

Saturn 1B rocket The Saturn-1B ("one-bee") rocket carried unmanned versions of the Apollo Command and Service Module and (separately) the Lunar Module into orbit to test their designs in 1966-68. The first stage was built by Chrysler Corporation and the second stage by Douglas Aircraft. The second stage, known as the S-IVB, was also used on the much more powerful Saturn V rocket.

Saturn V rocket The Saturn V carried the Apollo spacecraft to the Moon. Its first stage was powered by five Rocketdyne F-1 engines burning rocket grade kerosene in liquid oxygen and built by Boeing. The second stage was comprised of five Rocketdyne J-2 engines using liquid hydrogen and liquid oxygen as propellant and built by North American Aviation. The third stage, called the S-IVB, used a single J-2 engine and was built by Douglas Aircraft. Fourteen Saturn Vs were launched, all successfully. Three are on display in Huntsville, Houston and Cape Canaveral. See Chapter 2.

Service Module A Service Module is frequently split off from other parts of a spacecraft (robotic or manned) and contains the generic resources and equipment needed by every spacecraft, such as generating and storing electrical power, radio communications, propulsion, pointing and guidance equipment, temperature control, oxygen and carbon dioxide scrubbing (human missions), etc.

S-IVB The S-IVB ("ess-four-bee") was the third stage of the Saturn V and the second stage of the Saturn 1B. See Chapter 2.

Space Launch System (SLS) NASA's new heavy-lift rocket is called the Space Launch System (SLS). Its development is a key feature of the proposed Lunar Gateway program. The Orion spacecraft is being developed in parallel to be carried into space by the SLS. See Chapter 10.

Soyuz The word Soyuz is confusingly used to describe the rocket of that name and the manned spacecraft that it carries into space. Developed in the 1960s by the Korolev group, the Soyuz spacecraft was designed to carry three people into space for stays up to two weeks long and capable of maneuvering in space and enabling the crew to undertake space walks. With many gradual improvements the spacecraft continues to be the mainstay of the Russian human space program. The Soyuz rocket was derived from the R-7 long-range missile. There have been more than 1,700 launches of the various forms of the Soyuz rocket, and it continues to be used for human and robotic spaceflights. The Soyuz rocket is usually launched from Baikonur in Kazakhstan, but one version can be launched from Europe's launch site at Kourou in French Guiana in South America, which is close to the equator and thus increases the weight of payload it can place in orbit due to the more rapid Earth rotation at the equator.

Space walk Wearing a spacesuit, an astronaut is able to leave the spacecraft, i.e., to take a "space walk." NASA's name for this is "extra-vehicular activity (EVA)". If the space walk takes place on the Moon instead of in outer space it is commonly referred to as a Moon walk. The spacesuit acts as a miniature spacecraft, although with enough oxygen, electrical power, etc., for only a few hours.

SpaceX Founded by Elon Musk in 2002, SpaceX is one of the most successful space rocket companies in the world. Its Falcon 9 rocket, whose first stage can be recovered and reused, launches commercial, military and NASA spacecraft. Its Falcon Heavy rocket is the most powerful in the world. It is headquartered in Hawthorne (southwestern Los Angeles), California. See Chapter 10.

Sputnik Meaning "satellite" in Russian, the 184 pounds (83.6 kg) *Sputnik-1* was the world's first artificial satellite when launched by the Soviet Union on October 4, 1957. Launched a month later and weighing half a ton, *Sputnik-2* had a pressurized cabin and carried the dog Laika into space. A total of ten satellites launched by the Soviet Union were given the Sputnik designation, most of which tested systems for carrying animals into space and safely returning them to Earth. They also performed pioneering scientific measurements of magnetic fields, charged particles, radiation, temperatures and other characteristics of space.

STK rocket Russia's proposed new heavy-lift rocket is known by the Russian acronym STK – see Chapter 12.

Strategic Arms Limitation Treaty (SALT) Throughout the 1960s the United States and the Soviet Union deployed thousands of nuclear weapons on missiles and aircraft, each country trying to ensure that it had more and better weapons than the other. Both countries realized that deploying better weapons simply triggered the same action from the other, and thus did not improve their security. Thanks to the reliable information supplied by surveillance satellites, both countries agreed to halt this wasteful exercise, resulting in the first Strategic Arms Limitation Treaty in 1972.

Super Heavy and Starship SpaceX Chief Executive Officer Elon Musk says that future SpaceX launches of all kinds will use this giant rocket. The first stage Super Heavy and the upper stage Starship will both be re-usable, unlike most other rockets that are used only once. Its payload to Earth orbit will be about 150 tons, and it is intended to carry humans to Mars and back. See more in Chapter 10.

Surveyor NASA landed five Surveyor probes on the Moon's surface in 1966-68 (two more failed on landing), primarily to demonstrate the principle of a soft landing. See Figs. 6.1 and 7.3.

Tiangong Two of China's Tiangong manned space stations have been launched into orbit around Earth. *Tiangong-1* was launched in 2011 and visited by Chinese astronauts on three occasions in 2011, 2012 and 2013 before being mothballed in 2016. It re-entered Earth's atmosphere in 2018, ending up in the south Pacific Ocean. *Tiangong-2* was launched in 2016 and visited for a month by two Chinese astronauts that same year. It has also been visited three times by an unmanned cargo vehicle.

Umbilical tower While on the launch pad, a rocket is attached via umbilical connectors to ground-based electrical power and other resources. In some cases the umbilical connections are housed in a tower alongside the vertical rocket – hence the term umbilical tower.

V2 rocket The first rocket capable of traveling hundreds of miles was the V2 developed in Germany during World War II by a team led by Wernher von Braun. Over 3,000 V2 rockets were launched with a range of about 200 miles (320 km).

Vertical/Vehicle Assembly Building The Vertical Assembly Building was built at Cape Canaveral to allow the Saturn V rocket to be assembled before being transported to the launch pad 3½ miles (6 km) away. Now called the Vehicle Assembly Building it has been used to assemble the space shuttle and soon the Space Launch System. See Chapter 5 and Fig. 3.1.

Voskhod Two Voskhod spacecraft were flown in 1964 and 1965, each establishing a space "first." *Voskhod-1* was the first space vehicle to hold three crew members. *Voskhod-2* carried two crew members and enabled one of them, Alexey Leonov, to perform the first space walk. Voskhod was derived from the Vostok spacecraft and diverted funding away from the development of the Soyuz spacecraft intended to replace Vostok. See Chapter 8.

Vostok Six Vostok one-person spacecraft carried Soviet cosmonauts into space, including the first human in space, Yuri Gagarin (April 12, 1961), and the first woman in space, Valentina Tereshkova (June 16, 1963). During the return to Earth, the cosmonaut had to eject from the capsule at an altitude of 23,000 feet (7 km) and descend to the ground by parachute while the capsule also descended by a separate parachute. The Zenit military reconnaissance satellite was a variation of the Vostok design, returning exposed film and its camera to Earth in place of a cosmonaut.

Zenit rocket The Zenit rocket was the last to be developed in the Soviet Union before it ceased to exist at the end of the Cold War. It was built in the Ukraine. A total of 84 have been launched, of which 74 were wholly or partially successful. About half of the launches took place between 1999 and 2014 through the commercial Sea Launch company. Launches ceased following the 2014 breakdown in relations between Russia and Ukraine. See Chapter 12. Note that the name Zenit was also used for a series of surveillance satellites.

Index

A
Abe, Shinzō, 203
Aerojet, 146
Airbus, 134, 166
Alarm (computer/program), 41–44, 51, 70
Albania, 47
Aldrin, Buzz, 18, 42–45, 48–50, 52, 53, 61, 68, 74, 75, 107, 154, 166, 170, 188
Amazon (company), 14, 75, 136, 167, 170
Angara rocket, 123, 195, 196
Antarctic, 121, 139, 164
Apollo 7, 105
Apollo 8, 20, 21, 30, 38, 39, 53–56, 60, 65, 76, 104, 105, 219
Apollo 10, 21, 32, 39, 40, 54, 60
Apollo 11
    images, 81
    landing, 32, 40, 55, 68, 70, 117
    launch, 20, 22, 52, 68, 70, 106, 116, 117, 172, 182, 211
Apollo 12, 68, 70, 73
Apollo 13, 21, 24, 69, 70
Apollo 14, 70, 217
Apollo 15, 27, 34, 43, 44, 72
Apollo 16, 21, 44, 72
Apollo 17, 56, 70, 71, 73, 110, 112, 121, 125, 153
Apollo program
    Apollo 1 fire, 65, 101
    budget, 62, 75, 127, 131
    lunar orbit rendezvous decision, 65
    management, 62, 78, 127
Apollo-Soyuz, 55, 127
Arctic, 139
Ariane rocket, 137, 200

Armstrong, Neil, 18, 19, 22–24, 26, 31, 32, 35, 38, 40–53, 61, 68, 74, 76, 106, 107, 117, 170, 173, 188, 212, 213
Ascension island, 60
Atlas (rocket), 15, 123, 175
Augustine, Norm, 128, 130
Augustine report, 128, 130
Australia, 60

B
Babakin, Georgiy, 102, 106, 109
Bacon, Kevin, 70
Baikonur, 88, 112, 115–117, 130, 192–194, 198
Baker, Bobby, 27
Bales, Stephen G, 42
Bangladesh, 205
Barmin, Vladimir, 87
Bean, Alan, 68, 69, 73
BE-4 engine, 172
Beijing, 182, 189, 190
Bell, David, 5
Bell Labs, 62
Belyayev, Pavel, 95
Beria, Lavrentiy, 90
Berlin Wall, 93, 94
Bezos, Jeff, 14, 75, 136, 148, 166–175, 212, 213
Big falcon rocket (BFR), 162
Blue origin, 136, 137, 144, 148, 166–175, 202, 210, 212, 213
Boeing, 15, 135, 136, 148, 158, 159, 205
Bogomolov, Aleksey, 87
Borman, Frank, 53, 55, 96
Brand, Vance, 55

Braun, Wernher von, 5, 6, 12, 13, 15–17, 33, 34, 58, 64, 85, 86, 137, 200
Bridenstine, Jim, 134, 135, 143–148, 153, 154, 179, 207
Bush, President George H.W. (senior), 134, 135, 143–149, 153, 154, 179, 207
Bush, President George, W., 1, 134, 143

C
California, 15, 40, 60, 75, 131, 155, 224
Canada
    Quebec, Canadian Space Agency, 150, 152
Canadarm, 203
Cape Canaveral, 3, 10, 28, 57–59, 148, 155, 158, 161, 173, 174, 185, 192, 198
Carbonite-2 satellite, 124
Castro, Fidel, 3
Centaur engine, 16
Cernan, Gene, 40, 56, 73, 121, 125, 210, 214
Chaffee, Roger, 28, 62, 65
Chandrayaan, 204
Chang'e
    1, 184, 186
    2, 182, 184, 187
    3, 149, 182, 187
    4, 82, 182–184, 188, 214
    5, 182, 184, 214
Chelomey, Vladimir, 92, 97, 99–102, 106
China
    launch sites, 185, 186, 194
    moon program, 134, 189, 214
ChinaSat-5A satellite, 132
Circumlunar, 99, 102–105, 165
Clementine, 139, 140
Clinton, Hilary, 1
Collins, Michael, 18, 20, 22, 24, 31, 35, 37, 50, 52, 53, 61, 106, 153, 188
Command and service module (CSM), 26–28, 30, 32, 34, 35, 38, 40, 42, 43, 45, 50–55, 57, 58, 60, 69, 70, 79
Committee of chief designers, 85
Conrad, Charles (Pete), 68, 69
CORONA satellite, 119, 129
Cosmodrome, 117, 193–199
Cosmonaut, 2, 83, 84, 93–95, 97, 99–103, 110, 112–114, 116, 118, 124, 192, 198
Crew dragon, 164
Crick, Francis, 131
Cuba, 3, 29, 75, 90, 91, 126
    Cuban Missile crisis, 75, 90
Cultural revolution, 178, 179

D
Debris (space), 179, 180
Delta rocket, 175
Deng Xiaoping, 188, 189
Disney, Walt, 6
Djibouti, 205
Dragon (spacecraft), 157
Draper, Professor Charles Stark, 30
Dryden, Hugh, 5
Duke, Charlie, 44, 72

E
Eagle symbol, 37
Eisenhower, President Dwight D (Ike), 2, 3, 62, 119, 120, 137
Elektron satellite, 98
Europe, 47, 123, 126, 128, 137, 143, 148, 152, 175, 194, 199–203, 214
    European Space Agency, 2, 88, 151, 152, 199
Extra-vehicular activity (EVA), 95
    first, 95

F
Falcon 9, 123, 155–157, 159–161, 163–165, 170, 171, 174, 175, 211
Falcon Heavy, 123, 124, 136, 153, 157, 159–164, 166, 169, 173, 175, 211–213
Farside (of Moon), 2, 51, 54, 79, 82, 83, 105, 140, 141, 151, 152, 182, 184, 187, 214
Federation spacecraft, 198
F-1 engine, 11, 12, 14–16, 19, 66
Feniks rocket, 197
Fischer, Jack, 150, 151
Florida, 3, 19, 29, 58, 60, 70, 73, 75, 161, 169
Fobos-Grunt probe, 207

G
Gaganyaan, 204
Gagarin, Yuri, 2–6, 11, 62, 83, 84, 89–91, 93–96, 98, 119, 126, 129, 130, 194
Garver, Lori, 171
Gemini (space program), 57, 95, 96, 120
Geostationary orbit, 123, 124, 132, 133, 153
Geosynchronous space situational awareness program (GSSAP) satellites, 133
Germany, 5, 12, 85, 90, 94, 129, 137, 150–152, 200
Gerstenmaier, William H, 135
Glennan, T Keith, 62
Glenn, John, 70, 75, 169
Glushko, Valentin, 87, 98, 195

Google, 134
Gordon, Richard, 68
Gore, Al, 1
GR-1, 98
　　See also Orbital bombardment system
Grand canary, 60
Gravity, 9, 21, 24–26, 32, 34, 41, 44, 45, 53, 58, 60, 71, 72, 79, 88, 95, 96, 107, 168
　　Moon's, 26, 32, 34, 41, 44, 45, 53, 58, 71, 72, 106
Grissom, Gus, 27
GSLV rocket, 204
GSSAP, *see* Geosynchronous space situational awareness program (GSSAP) satellites
Guam, 60

H
Hadfield, Chris, 203
Haise, Fred, 21
Hanks, Tom, 70
Hasselblad, 48
Hawaii, 55, 60
Henry III, King, 48
Hermes spacecraft, 201
H-IIB rocket, 123
Hiten space probe, 111
Holbein, Hans, 48
Houston TX, 7, 30–32, 41, 43, 54, 58, 60, 71, 125, 127, 188
Hubble space telescope, 181
Huntsville AL, 5, 6, 14, 17, 63

I
IBM, 17, 60, 131
ICBM, *see* Inter-Continental Ballistic Missile (ICBM)
India, 77, 104, 123, 143, 175, 187, 189, 194, 199, 204–208, 214
Indian Space Research Organisation (ISRO), 204, 205, 207
Intel, 125
Inter-Continental Ballistic Missile (ICBM), 65, 88, 98, 119, 129
International Space Station, 88, 122, 124, 130, 140, 145, 148, 150, 174, 199
Irwin, Jim, 72
ISRO, *see* Indian Space Research Organisation (ISRO)

J
Japan, 123, 141, 148, 175, 178, 179, 194, 199, 201–203, 205, 214
J-2 engine, 17, 24
Jiuquan, 186
Johnson, Lyndon B (LBJ), 5, 7, 62, 129
Juno rover, 150, 152

K
Kaguya space probe, 77
Kamanin, Lt. General Nikolai, 103
Kazakhstan, 108, 192–194, 198, 199
KBKhA, 198
Kelly, Scott, 165
Kennedy, John F. (President), 1–8, 46, 47, 61, 74, 75, 79, 89, 98, 101, 119, 120, 127, 137, 210
Kennedy, Robert (Bobby), 4
Kennedy, Senator Ted, 47
Kepler, Johannes, 38
Kerr, Senator Robert S, 27
Khrushchev, Nikita, 88, 90, 92, 130
Khrushchev, Sergei, 92
Komarov, Vladimir, 103
Kopechne, Mary Jo, 47
Kopra, Tim, 89
Korea, 47, 126, 203
　　North, 47
Korolev, Sergei, 84–87, 92, 195
　　death, 100
Kozlov, Frol, 130
Kuznetsov, Viktor, 87

L
L3, 99-101, 112
Laika (dog), 2
Lalithambika, V.R., 207
Landing sites (on Moon), 40, 68, 79, 108, 109, 187
Launch Escape System, 20–22
Launius, Roger, 74
L1 circumlunar spacecraft, 104
Leonov, Alexei, 93–95
Liwei, Yang, 179, 180
LK spacecraft, 99, 100
LOK, 113, 114
Longjiang space probe, 77
Long March, 188
　　-4/-5/-6, 123, 132, 184–186
Lotos (satellite), 195
Lovell, Jim, 70, 96
Low, George, 5

Luna
-15, 52, 53, 107
-16, 85, 107, 108
-17, 110
-20, 85, 108
-21, 110, 112
-24, 85, 108, 111
sample return, 102, 109, 125, 184
Lunar gateway, 135, 143–145, 147, 149, 150, 152–154, 196, 201–203, 206
Lunar module, 24, 25, 27, 31, 32, 34, 35, 37–43, 45, 47–53, 57, 58, 60, 69, 72, 73, 78, 97, 105, 112, 113, 120, 121, 188
Lunar orbiter spacecraft, 57, 106
Lunar reconnaissance orbiter, 77
Lunar-X prize, 134
Lunokhod, 106, 125, 149, 186

M

Macron, Emmanuel (President), 201, 204
Maezawa, Yusaku, 165
Malenchenko, Yuri, 88
Mangalyaan, 204
Manned orbiting laboratory (MOL), 138, 139
Mares (lunar), *see* Seas (lunar)
Mars (planet), 60, 80, 84, 88, 98, 102, 128, 131, 134, 135, 140, 143, 152–153, 161, 163–166, 173, 175, 181, 186, 201, 204, 207, 211, 212
Marshall Space Flight Center, 5
Martin company, 26
Mascon, 41
Massachusetts Institute of Technology (MIT), 30
Mattingly, T Kenneth (Ken), 21
McNamara, Robert, 63, 137
Mercury (space program), 130
Merlin engine, 160, 162, 163
Methane, 163, 164, 171
Minuteman ICBM, 65
Mishin, Vasily, 102
Mission control, 23, 24, 30, 32, 40–43, 50, 52, 54–56, 60, 61, 68, 70, 71, 110, 125, 150, 188
Mitchell, Edgar, 70
Modi, Narendra, 204–207
Mogensen, Andreas, 151
Molniya satellite, 98
Moon dust/rock, 47, 48, 50, 53, 55, 68, 70, 79, 80, 102, 107, 109, 182, 214
cost, 109
Moon express, 134
Moon farside, 105, 140
Moore, Gordon E, 125

Moore's law, 125, 126
Moscow, 2, 47, 53, 90, 91, 94, 103, 104, 117, 140, 193
Mowry, Clay, 172
Mueller, George, 63, 64
Muller, Paul, 40
Musk, Elon, 136, 146–148, 154–155, 159, 162, 166, 168, 170, 211, 212

N

N1, 53, 97–101, 111–119, 123, 136
National Advisory Committee for Aeronautics (NACA), 64
National Aeronautics and Space Administration (NASA), 4–9, 11–16, 19, 20, 22, 23, 26–28, 31, 34, 35, 37–39, 42, 44, 46, 48, 49, 51, 52, 54, 58–60, 62, 63, 66, 68, 69, 71, 72, 74–78, 80, 82, 88, 93–96, 99, 101, 103, 105, 107, 109, 113–116, 120, 122, 127, 128, 131, 132, 134–137, 139–141, 143–154, 156–158, 161, 164, 165, 170, 171, 173, 175, 179, 183, 187, 201, 202, 205, 207, 208, 210, 211, 213
inspector general, 148
National Research Council, 81, 82
National Space Council, 135
Naval Research Laboratory, 139
Nedelin, Marshall Mitrofan, 130
New Armstrong, 173, 212, 213
New Glenn, 136, 169–173, 175, 212
New Shepard, 167–170, 172, 174, 175
Newton, Isaac, 38
Nigeria, 134
Nixon, President Richard, 1, 48, 55, 58, 139
North American Aviation, 15, 26–28, 62
North Carolina, 62
Nurse, Paul, 75

O

Obama, Barack (President), 158
Orbital ATK, 135
Orbital bombardment system, 98
See also GR-1
Origin (of Moon), 79
Orion spacecraft, 154, 201, 202

P

Pakistan, 205
Pauling, Linus, 131
Peake, Tim, 88
Pearl Harbor, 64

Pence, Vice President Mike, 135, 153
People's Liberation Army, 179, 180
Pesquet, Thomas, 201
Phillips, General Sam, 65
Pilyugin, Nikolay, 87
Plesetsk, 193
Polaris missile, 30
Power and Propulsion Element, 135
Pratt & Whitney, 17
Proton rocket (UR-500), 92, 99, 101, 102
Putin, President Vladimir, 140, 190, 192–197, 199, 201

Q
Queqiao, 182, 184, 187

R
Ranger, 186
Raptor engine, 163
Rayburn, Sam T, 7
RD-270 rocket engine, 98
Rendezvous radar, 43, 45, 51
Renmin University, 189
Resource Prospector space probe, 134
RKTs Progress, 198
R-7 missile, 129
R-9 missile, 92, 98
Rocketdyne, 11, 15–17
Roosa, Stuart, 70
Roscosmos, 196
RS-25 engine, 146
RS-16 missile, 92
RT-1 missile, 98
RT-2 missile, 98
Russia, 88, 89, 123, 132, 133, 148, 175, 179, 190, 192–208, 214
   launch sites, 193, 194
Ryazanskiy, Mikhail, 87

S
Salyut space station, 84
Saturn 1 (rocket), 9
Saturn V
   first stage, 11, 167
   second stage, 11, 15
   third stage, 11, 24
Schirra, Wally, 96
Schmitt, Harrison (Jack), 73, 112
Science (of Moon), 5, 72, 75, 76
Scott, Dave, 72
Seas (lunar), 40, 185

Service Module
   CSM, 26–32, 55, 69
   Orion, 201, 202
Shambaugh, David, 189
Shea, Joe, 65
Shenzhou, 181, 189, 205
Shepard, Alan, 3, 4, 6, 70, 130, 168
Shi Jian satellite, 132
Shotwell, Gwynne, 160
S-IVB, 17
Sjogren, Bill, 40, 41
Slayton, Deke, 55
Sorensen, Teddy, 5
South pole Aitken, 182, 183
Soviet Union
   circumlunar, 105
   manned, 96, 97, 103, 105
   sample return, 81, 82
Soyuz rocket, 119, 194
Soyuz spacecraft, 93, 101, 113, 198, 199
Space Launch System (SLS), 136, 144–146, 149, 157, 161–165, 169, 170, 175, 201, 213
Space Shuttle, 55, 119, 124, 137, 146, 147, 149, 153, 161, 163, 171, 192, 200, 203
SpaceX, 135–137, 144, 148, 154–175, 197, 198, 201, 202, 205, 210–213
Spain, 60
Sputnik, 3, 4, 11, 62, 89, 90, 98, 119, 128–130, 162
Stafford, Tom, 21, 55, 96
Stalin, Josef, 87, 88, 90, 128
Starship, 136, 162-166, 173, 211, 212
STK rocket, 198
St Louis MO, 63
Strategic Arms Limitation Treaty (SALT-1), 98
Submarine Launched Ballistic Missiles (SLBMs), 129
Sunkar rocket, 197
Super Heavy, 136, 157, 162-166, 173, 185, 196-199
Surveyor-3, 68, 69
Swigert, John (Rusty), 69

T
Taiwan, 126, 132
Telstar satellite, 124
Tereshkova, Valentina, 94
Tesla, 136, 157, 161
Texas, 1, 7, 75, 167
Tiangong, 181
Tianhe, 181, 184
Titan rocket, 92
Titov, German, 93, 94
Toutatis (asteroid), 187

Tracy's Rock, 73
Tranquility Base, 43, 51, 53
Trump, Donald (President), 1, 134, 135, 143, 146, 154, 180, 195, 208
TRW, 9, 30, 34, 41, 56, 60, 63
Tselina (satellite), 195
Tsiolkovsky, Konstantin, 33, 84, 88
Tupolev, Andrey, 87
Turkey (country), 90

U
Ukraine, 110, 192, 195
United Launch Alliance, 147, 171–175
United Nations, 74
UR-500 rocket, *see* Proton rocket (UR-500)
UR-700 rocket, 99
US Air Force, 15, 65, 133, 138, 162
USSR Union of Soviet Socialist Republics, *see* Soviet Union
Ustinov, Dmitriy, 130

W
Wang Wen, 189
Watson, James, 131
Watson, Thomas J Jr, 131
Webb, James, 5, 26, 62, 74, 75
Wei Fenghe, 206
Wenchang, 186
White, Ed, 28
Wiesner, Jerome, 5
Williams, Sunita, 207, 208
Worden, Al, 72

X
Xi Jinping, 189, 190

Y
Yangel, Mikhail, 92
Yang, Liu, 179, 180
York, Herbert, 129
Young, John, 72
Yutu rover, 149, 187

Z
Zak, Anatoly, 199
Zedong, Mao, 179, 190
Zenit rocket, 195–197
Zenit satellite, 98, 129
Zond space probes, 104, 105